4 Springer Series in Solid-State Sciences
Edited by Peter Fulde

Springer Series in Solid-State Sciences

Editors: M. Cardona P. Fulde H.-J. Queisser

Inelastic Electron Tunneling Spectroscopy

Proceedings of the International Conference,
and Symposium on Electron Tunneling
University of Missouri-Columbia, USA,
May 25–27, 1977

Editor T. Wolfram

With 126 Figures

Springer-Verlag Berlin Heidelberg New York 1978

Professor Thomas Wolfram, PhD

University of Missouri-Columbia, College of Arts and Science, Department of Physics,
223 Physics Building, Columbia, MO 65201, USA

Editors:
Professor Dr. Manuel Cardona
Professor Dr. Peter Fulde
Professor Dr. Hans-Joachim Queisser

Max-Planck-Institut für Festkörperforschung
Büsnauer Strasse 171, D-7000 Stuttgart 80, Fed. Rep. of Germany

The Conference and Symposium were supported by

The National Science Foundation
The Office of Naval Research
U. S. Energy Research and Development Administration

Ford Motor Company
Rockwell International Corporation

The University of Missouri-Columbia, Graduate School
The University of Missouri-Columbia, Department of Physics
The University of Missouri-Columbia, College of Arts and Sciences

Dr. Paul E. Basye

ISBN 3-540-08691-9 Springer-Verlag Berlin Heidelberg New York
ISBN 0-387-08691-9 Springer-Verlag New York Heidelberg Berlin

Offset printing and bookbinding: Zechnersche Buchdruckerei, Speyer.
2153/3130-543210

Preface

Inelastic Electron Tunneling Spectroscopy, or IETS, provides a unique technique for electronically monitoring the vibrational modes of molecules adsorbed on a metal oxide surface. Since the discovery of the phenomena by JAKLEVIC and LAMBE in 1966, IETS has been developed by a number of scientists as a method for studying the surface chemistry of molecular species adsorbed on aluminum oxide. Recent applications of IETS include investigations of physical and chemical adsorption of hydrocarbons, studies of catalysis by metal particles, detection and identification of trace substances in air and water, and studies of biological molecules and electron damage to such molecules. IETS has been employed to investigate adhesive materials, and studies are currently in progress to investigate corrosion species and corrosion inhibitors on aluminum and its alloys. Electronic transitions of molecules have also been studied by IETS.

The recent development of the "external doping" technique, whereby molecular species can be introduced into fabricated tunnel junctions, opens the door for a vast new array of surface chemical studies by IETS. IETS is rapidly becoming an important tool for the study of surface and interface phenomena.

In addition to its role in surface studies, inelastic tunneling has proved extremely valuable for the study of the electronic properties of thin metallic films, and the recent discovery of light emission from inelastic tunneling promises to be of some importance in the area of device physics.

The Conference and Symposium provided a forum for scientists actively engaged in IETS to discuss recent progress and it also served to introduce IETS to a broader community of scientists.

The International Conference on Inelastic Electron Tunneling Spectroscopy was held at the Department of Physics of the University of Missouri, Columbia, Missouri from May 25 to 27, 1977.

The Conference and Symposium was supported by grants from the National Science Foundation, the Office of Naval Research, and the United States Energy Research and Development Administration. Financial support was also supplied by the Ford Motor Company and the Rockwell International Corporation. We wish to acknowledge support from the University of Missouri, Graduate School, College of Arts and Science and the Department of Physics. Special acknowledgment is also given to Dr. Paul E. Basye for financial assistance and continued support of the activities of the Department of Physics at the University of Missouri.

About 200 scientists from the United States, Canada, France, England and North Ireland attended the Conference and Symposium.

The Conference, held on May 25 and 26, provided an extensive review of the technique of IETS and of the applications of IETS to a variety of different areas of study. The Symposium on Electron Tunneling was held on May 27 and included a review of superconducting tunneling phenomena together with recent advances in IETS.

All of the papers presented at the Conference and Symposium were invited by the Conference and Symposium Organizing and Program Committee. Nearly all of these papers are presented in full length in the proceedings. A few of the papers, which were submitted elsewhere for publication, appear here only in abstract form.

The papers presented at these proceedings follow closely the order in which they were given in the Conference and Symposium, however, some reordering has been made for the sake of continuity.

The Members of the Organizing and Program Committee are listed on p. 239, as well as the original programs on pp. 240 and 242. I am very grateful to them for their assistance in organizing, and their participation in the Conference and Symposium. Finally, I would like to thank Mrs. Beverly Huckaba for her assistance and support in all phases of the organizing, soliciting and logistics of the Conference and Symposium.

December 1977   Thomas Wolfram

Contents

I. Review of Inelastic Electron Tunneling

Inelastic Electron Tunneling Spectroscopy (IETS) – Past and Future

R.C. Jaklevic

Research Staff, Ford Motor Company
Dearborn, MI 48121, USA

ABSTRACT

An outline of the history of inelastic tunneling must begin with the experiments of IVAN GIAEVER with superconducting metals which established the physical reality of the metal-insulator-metal tunneling method. Early attempts to observe electrode band structure effects at higher voltages were not successful but led to the discovery of inelastic tunneling spectroscopy of organic molecules, oxides, and electrode phonons. The study of a variety of organic species has been an active area of experimentation. Observations of electronic transitions, plasmons quantum light emission are of current interest. Possible future developments will be discussed.

In this talk I will briefly recount those events in the past which involve the development of tunneling, and the experiments which have contributed to our present understanding of IETS. Most of the developments of recent years will be treated in much greater detail during this Conference. Tunneling, of course, is as old as quantum mechanics [1]. It occurs as a result of the fact that particles and their wave functions can penetrate a potential barrier while classical mechanics forbade them from doing this. The effect becomes important if the potential barriers are very narrow, of the order of several wavelengths wide. Among the first topics which were treated in quantum mechanics were the field ionization of atoms with electric fields and the emission of alpha particles from heavy nuclei. Concerning electron tunneling in solids, however, most of the early work was in the area of theory. FOWLER and NORDHEIM in 1928 showed that if one applied a very large electric field to the surface of a metal, electrons could tunnel out into vacuum. Before 1940, the simple theory was done for tunneling from one metal to another through either a vacuum or insulating gap (SUMMERFIELD and BETHE, FRENKEL). Another similar calculation due to ZENER in 1943 showed that internal field emission can be caused by application of a very large electric field inside of an insulator in which electrons tunnel from one band to another. As far as solid state tunneling was concerned, however, not much further progress was made until about 1960, primarily because of the lack of suitable experiments. There were some early attempts to use tunneling to explain the existing facts known about rectification at metal-semiconductor contacts but the tunneling model predicted the wrong direction for current flow. It wasn't until about 1960 that experimenters were able to show conclusively that there are situations in the solid state where electron tunneling occurs and that the effect can be used as a tool for studying other solid state phenomena.

The first of these experiments was the invention of the ESAKI diode in 1957. ESAKI's experiments [2] involved a semiconductor p-n junction diode where current flow is caused by diffusion of electrons and holes from one side of the junction to the other. EASKI showed that if the doping gradient of a p-n junction is made very narrow (Fig. 1), the wave functions of electrons from one side could extend through the forbidden gap over into states

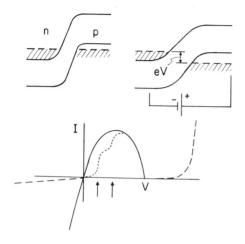

Fig.1 Schematic of energy bands of ESAKI diode where the doping gradient extends over about 100Å. The usual p-n diode diffusion characteristics (broken line) go over into the tunneling behavior (solid line). Phonon effects (dotted line) show up at low temperatures

on the other side. Tunneling can therefore occur and large current flows near zero bias with a forward characteristic in the shape of an inverted parabola. This behavior is completely different from an ordinary p-n diode and the effect is so striking and unique it can only be explained by tunneling. The early workers in this field cooled these diodes and observed structure in the tunneling current occurring at voltages corresponding to phonon energies [2]. For this kind of tunneling, called interband tunneling, it turns out that there are momentum selection rules very similar to the direct and indirect selection rules of optical absorption in solids. In the case of Ge and Si, the transitions are indirect and phonon emission is involved in tunneling. So this "phonon assisted tunneling" in a sense was the first example of inelastic tunneling.

The next experiments were those of IVAR GIAEVER and JOHN FISHER [3] at General Electric in 1960. They were the first to make a careful attempt to fabricate a tunnel junction of the type metal-insulator-metal by the use of vacuum evaporation techniques. The procedure (Fig. 2) has not changed much since then; the first film, say aluminum, is deposited by vacuum evaporation and then is oxidized to about 20 to 30Å thickness so that tunneling can occur. Then the top metal electrode is deposited in the form of a cross strip. A current is applied to two terminals and the voltage across the oxide is measured at the other two terminals. They obtained current-voltage curves which were very reasonable and apparently were tunneling because their behavior agreed with the predictions of the simple theory. First, the conductance dI/dV had a roughly parabolic shape (Fig. 3). At low voltages the resistance is constant and they found that this resistance could be varied over a large range by changing the thickness of the aluminum oxide. This resistance has an exponential dependence on oxide thickness as the simple theory predicts. Furthermore, they found that the resistance is not strongly dependent on temperature. This is another characteristic of tunneling due to the fact that metal electrons occupy a broad band of allowed energy states.

3

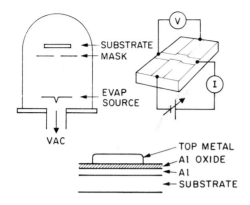

Fig.2 Schematic showing some experimental details of metal-metal oxide-metal tunneling. The metal films are evaporated and the Al oxide is grown to a thickness of about 25Å by exposure to oxygen or air

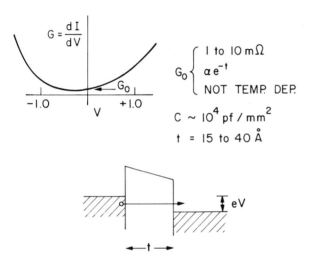

$$G = \frac{dI}{dV}$$

$$G_0 \begin{cases} 1 \text{ to } 10 \, m\Omega \\ \alpha e^{-t} \\ \text{NOT TEMP. DEP.} \end{cases}$$

$$C \sim 10^4 \, pf \, / \, mm^2$$

$$t = 15 \text{ to } 40 \, \overset{\circ}{A}$$

Fig.3 Characteristics of M-I-M tunneling showing the inverted parabolic shape of conductance versus voltage. All of the potential difference of the applied voltage is supported by the oxide layer

Also, they found that the capacity of the junctions was very high (about .02 µf/mm). While these facts could easily be explained by the simple theory, there are enough adjustable parameters available when fitting observed curves to theoretical ones that it proved almost impossible to learn anything about the details of real tunnel junctions simply by examining I-V curves at room temperature.

The next experiment was GIAEVER's discovery of the superconducting density of states in tunneling [4]. He found that if one of the electrodes was super-conducting, there appears a very characteristic structure within a few milli-volts of zero bias which is directly related to the density of states of the superconducting Pb (Fig. 4). This was the first direct measurement of this

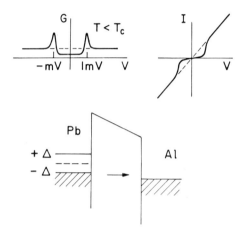

Fig.4 Schematic of superconducting effects seen in M-I-M tunneling. At very low temperatures T $\ll$ T$_c$ the conductance is nearly proportional to the super-conducting density of states. Note the large decrease in conductance at zero bias due to superconductivity

important property of superconductors. This whole area of experimentation has become a field in itself. For a very practical reason, however, these effects play an important role in IETS as well. Each time an attempt is made to fabricate a tunnel junction, there will be some kind of current flow through the junction and some kind of data will be gotten, whether related to tunneling or not. This is because other types of current flow mechanisms such as tiny metallic bridges, two step tunneling or diffusion through the oxide may be operating. A common first test is to look for the existence of the superconducting structure and to check whether it is of the proper shape, etc. This method of checking has often been emphasized and is a par-amount reason why these early experiments were of importance to IETS.

Following these early works, tunneling followed a variety of paths in-volving other techniques [5]. The Josephson effect, whereby supercurrents can tunnel through a barrier, has by now become a field on its own [6]. Also there has been interest in another type of tunnel device, the Schottky diode junction. In these experiments a piece of bulk degenerate semiconductor is in contact with a metal and the resulting space charge barrier is so narrow that tunneling can occur. A variety of effects have been studied with these devices including magnetic Landau levels and scattering from localized spin states. So far no molecular IETS has been observed for these devices, a puzzle which is not yet understood.

The early motivation for studying nonsuperconducting effects in M-I-M junction came from hopes that one might observe metal band structure effects. Our reasons also included the possibility of observing spin density waves in Cr. As it happened, our search for new structure at higher voltages led to the discovery of IETS of organic molecules [7]. Experimental study of band structure by tunneling was to wait for a number of years.

The discovery of inelastic electron excitation of molecules resulted from an experimental examination of junctions in the range up to 0.5 volts. If one looks closely at the I-V curve for an M-I-M junction cooled to 4.2° K, the otherwise smooth curve shows small increases in the slope at certain voltages in both bias directions (Fig. 5). In dI/dV there are small steps

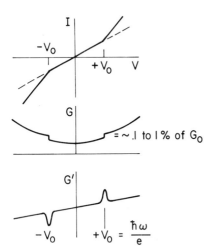

Fig.5 Schematic of inelastic tunneling effects seen in M-I-M junction with vibrational mode excited at $V_0 = \hbar\omega/e$

of the order of .1 to 1% of the conductance. Therefore these effects cannot be seen directly on an oscilloscope or curve tracer. If we go to d^2I/dV^2 peaks appear and now the data begin to look like spectra. These peaks correspond to the energy associated with the vibrational energy of the molecular species in the barrier.

An important development in experimentation is worth mentioning here in connection with the value of modulation techniques in tunneling. In many areas of instrumentation, there has been a gradual adoption of a.c. techniques which allow the detection of small structures in the presence of a large background. Without these methods, early experimenters who studied M-I-M junctions by looking at I versus V had little chance of seeing IETS. With modulation methods, these observations become sensitive and direct. The method (Fig. 6) essentially involves the application of a small a.c. signal on top

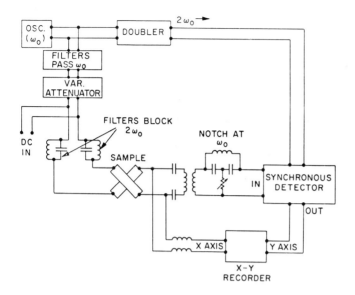

Fig.6 Schematic showing a second harmonic method for detecting the second
derivative d^2I/dV^2. The slowly varying d.c. voltage corresponds to energy
while the second harmonic component of the a.c. modulation voltage appearing
across the junction is proportional to d^2I/dV^2

of the d.c. bias voltage and observing the first or second harmonic of the
response of the junction. The second harmonic corresponds to the now familiar
second derivative IETS spectrum. Background effects are greatly reduced and
the limit of detectability is determined by the noise generated by the junction itself. Ideally it is a shot noise device so that the noise will increase with current. At low temperatures, a voltage of 50 mV is required
for the noise of a tunnel junction to equal that of a room temperature
Johnson noise resistor.

A typical IETS spectrum might look something like the fictitious one
shown in Fig. 7. It covers an energy range of about 20 to 500 meV, the
range of the optical near infrared, corresponding to 100 to 4000 cm^{-1}. In
this range we see typically hydroxyl modes OH, CH stretching and bending
modes, and other groups such as NH, CN, CO and so on. Just as the case of
IR and Raman spectroscopy we find the characteristic vibrations in groups
which can identify the molecule. Numerous spectra of this type have been
obtained and will be seen at this conference.

Much of our early experiments concentrated on learning the important
characteristics of IETS. The first one, of course, is the fact that the position of peaks of d^2I/dV^2 are directly related to the frequency of oscilla-
tion of the molecules by $eV = \hbar\omega$. The molecules comprise about one monolayer
or less which translates to a total number of about 10^{13} or 10^{14} and a detectability lower limit of about 10^{10}. So, in absolute numbers of molecules, the

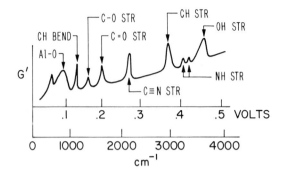

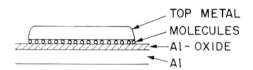

Fig.7 Sketch of an IETS spectrum showing a number of different molecular
vibration peaks. The cross sectional sketch of a tunnel junction shows the
molecular layer adsorbed on the oxide-metal interface

technique is much more sensitive than usual IR methods by a factor of a hundred thousand or more. The line width of an IETS peak was found to be increased with temperature with optimum resolution achieved at 4° K or below.
This temperature broadening provided an important clue toward understanding
the mechanism of IETS. The observed frequencies of the molecular vibrations
are not found to be appreciably changed as a result of being buried inside
of the tunnel junction. Of course, there are exceptions in those cases where
chemisorption occurs, such as the adsorption of an organic acid. Additional
inelastic interactions were observed for the case of oxide phonons and metal
electrode phonons [5].

All of these experimental facts led to the interpretation of IETS by a
simple theoretical model. The observed temperature dependence indicates that
the electrons involved are originating near the Fermi levels of the metal
electrodes. Since the interaction of tunneling electrons with local vibrations causes an increase in tunnel conductance, a fairly simple picture of
this effect emerges (Fig. 8). The voltage across the tunnel junction is
the difference in electron volts between the two Fermi levels. If the tunneling electron is to give up energy to the vibrator it must be able to find
an empty state after it tunnels across. This means that the voltage must be
$\hbar\omega/e$ or greater. The inelastic transition becomes possible and, as the voltage is increased, a larger number of electrons can tunnel by this process.
Therefore, a new channel for tunneling has been turned on by the inelastic
interaction.

8

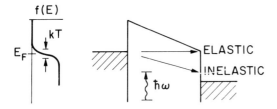

Fig.8 Sketch of the energy levels of a junction showing the tunneling transitions (elastic and inelastic) possible. A change in resistance occurs when the inelastic channel is turned on. The sharpness of the turning on is determined by the temperature broadening of the Fermi edge in the metal electrodes

A simple theoretical calculation based on this model was worked out by SCALAPINO and MARCUS [8] in which the electron-molecule interaction was added on to the simple one electron tunneling theory. In their model (Fig. 9) the

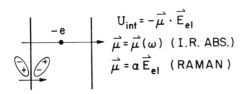

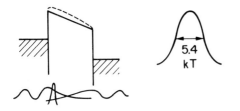

Fig.9 Electron-molecule interaction model used to calculate IETS current. The interaction energy U_{int} results from the electron-dipole and electron-induced dipole interactions. The line width due to temperature broadening is 5.4 kT

electric dipole moment of the molecule, including its image, was placed near one of the metal surfaces and the energy of interaction U_{int} of the molecule with the tunneling electron was calculated. The dipole moment varies with the same frequency as the vibration so that the interaction was viewed as a modulation of the barrier height by the molecular vibration. They found that there is an increase in the tunneling conductance of about the same size as was observed. Also, the selection rule involved turns out to be the same as are involved in infrared absorption, the electric dipole selection rules. Temperature broadening naturally comes out of the theory as a folding of the Fermi functions of the two metal electrodes. A definite line width of 5.4 kT was predicted in agreement with observations.

There is another way for a tunneling electron to interact with a molecule. Because the bonds between atoms are polarizable the field of the tunneling electron can induce a dipole moment in the molecule and therefore there exists an energy of interaction. When that term is taken into account one obtains a contribution to the electron current due to processes obeying the Raman selection rules. What happens is that electrons really have no respect for optical selection rules and are interacting with all of the charges in the molecule. More accurate and complete theories of this interaction have been done and it is found that almost any mode, regardless of symmetry or how forbidden it may be in optical experiments, is detectable in IETS experiments.

During the past ten years a number of experiments and theoretical developments have been carried out on the subject of IETS [9]. A whole variety of organic molecules including acids, bases, ring compounds and biologically interesting materials have been studied. Studies have been made to apply the method to trace analysis. Also oxide and normal metal phonons have been studied [5]. The temperature dependence of the line width has been measured with care and agreement with theory has been good. The effect of various electrode metals has been studied as far as they affect the intensity and position of the spectral lines. Some metal overlays give much more intense spectra (e.g. Pb) and some very weak spectra (e.g. Al). As it turns out, the larger the overlay atom the less it interferes with the molecular layer. There are also slight shifts in frequency due to the interaction of a vibrating dipole with the metal electrode. The selection rule effects have been studied by using molecules which have modes of known symmetry and selection rules and showed that indeed Raman allowed modes are observable in IETS. Orientational effects resulting from the interaction of a dipole with its image have been studied. These give a large effect for dipole perpendicular to the metal electrode and a much smaller effect for those moments parallel to the electrode. Chemical changes can cause large changes in the vibrational modes and have been observed for example in the case of acids whereby the organic species is chemisorbed on the oxide surface. Solvents, whose vapor pressure is high and whose molecules do not stick very well on the oxides, have been studied by low temperature deposition techniques. In this connection much effort has been spent learning how to deposit molecules on the oxide surface. Very low vapor pressure molecules have been deposited by evaporation or directly deposited on the oxide outside the vacuum system from a solution by dipping or spinning techniques. These include biologically interesting molecules which cannot be evaporated because they are very unstable with temperature. For the study of gases, work has been done in which a submonolayer amount of rhodium was deposited on the oxide and then exposed to carbon monoxide which chemisorbed onto the rhodium atom. The spectrum of the carboxyl group was then seen. High energy electrons have been projected through the thin film structure and the damage to the molecules contained therein was observed in the IETS. Another area of work involves the observation of electron interaction with localized spins in the barrier region, analogous to Kondo scattering in metals. A recent new development is the discovery that molecules can be introduced into completed junctions by an infusion technique [10]. Junctions can be made and stored for long periods of time and then infused with molecules from the vapor, solid or liquid phase.

Another important area is that of electronic excitations. Effort extends
back to the early work in IETS in which it was noted that an electronic excitation should also be observable by IETS with even larger intensity. A
number of experiments have been reported involving the introduction of dyes
and other species such as rare earth compounds [9]. We will see some of
these results later in this Conference. Another class of electronic excitations is observed in a little different way, that is, by looking at light
emission from the tunnel diodes. In one experiment, the tunneling electron
excites a bulk plasmon in one metal electrode and the light radiated from
the de-excitation process is observed [11]. In another experiment, the tunneling electron can excite electromagnetic surface waves which then scatter
out into free space. This forms a tunable light source.

For purposes of comparison, Fig. 10 shows the energy scale over which

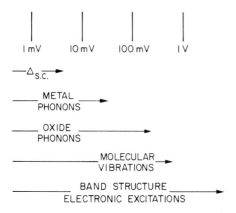

Fig.10 Tunneling energy scale

various tunneling experiments are conducted. As can be seen experiments have
been going to higher voltages for the case of band structure and electronic
excitation. In these instances limitations are encountered due to the breakdown problems in oxide (field strengths of the order of 10^7 v/cm).

There is a growing list of possible uses of IETS which have been envisioned
by various workers in the field. They concern surface chemistry, catalysts,
biological materials, metal-molecule interactions, trace impurity detection
and electronic excitations. The last are also important because of the possibility of using tunnel junction as practical light sources. Desirable
future improvements would include more reliable insulating barriers and tunnel devices which can withstand more rigorous chemical and temperature treat-

ments. Such properties would be necessary, for example, for a truly useful light source or for a reuseable molecular IETS device. New type of electrodes which have sharp tunneling structure due to narrow energy bands would allow spectra to be taken at room temperature instead of cryogenic temperatures. One possible source of such electrodes might be the recently developed techniques of molecular beam epitaxy.

References

1 C.B. Duke: Tunneling in Solids Academic Press (New York 1969). See the beginning chapters for a historical survey of tunneling
2 L. Esaki: Phys. Rev. 109, 603 (1957)
   L. Esaki: Rev. Mod. Phys. 46, 237 (1974)
3 J.C. Fisher, I. Giaever: J. Appl. Phys. 32, 172 (1961)
4 I. Giaever: Phys. Rev. Lett. 5, 147 (1960)
   I. Giaever: Rev. Mod. Phys. 46, 245 (1974)
5 For reviews of various non-superconducting tunneling experiments see E.L. Wolf, in Solid State Physics, ed. F. Seitz, D. Turnbull, H. Ehrenreich (Academic, New York, 1975),Vol. 30
6 B.D. Josephson: Rev. Mod. Phys. 46, 251 (1974)
7 R.C. Jaklevic, J. Lambe: Phys. Rev. Lett., 17, 1139 (1966). J. Lambe, R.C. Jaklevic: Phys. Rev. 165, 821 (1968)
8 D.J. Scalapino, S.M. Marcus: Phys. Rev. Lett. 18, 459 (1967)
9 For detailed reviews of IETS see P.K. Hansma: Physics Reports 30C, 146 (1977). R.G. Keil, L.P. Graham, K.P. Roenker: Appl. Spectroscopy 30, 1 (1976)
10 R.C. Jaklevic, M.R. Gaerttner: Appl. Phys. Lett. 30, 646 (1977)
11 Tien-Lai Hwang, S.E. Shwarz, R.K. Jain: Phys. Rev. Lett. 36, 379 (1976)
12 John Lambe, S.L. McCarthy: Phys. Rev. Lett. 37, 923 (1976)

Survey of Applications of Tunneling Spectroscopy [1]

P.K. Hansma [2]

Department of Physics, University of California
Santa Barbara, CA 93106, USA

ABSTRACT

Tunneling spectroscopy has unique capabilities for the vibrational spectroscopy of thin organic films. After an examination of its spectral range,
sensitivity, resolution, and selection rules, and a comparison of it with
other techniques for vibrational spectroscopy, I will discuss six areas of
application: (1) surface chemistry, (2) identification of biochemicals,
(3) trace substance detection, (4) adhesion, (5) radiation damage, and (6)
lubrication.

1. Introduction

Since we will have an entire session this afternoon on four applications of
tunneling spectroscopy, I will not need to cover these in any detail. Therefore I will use the first half of my time to discuss the more general question: Why does tunneling spectroscopy have applications? Specifically I
will discuss the capabilities of tunneling spectroscopy and compare it to
other techniques for vibrational spectroscopy. From this discussion I hope
it will become clearer why tunneling spectroscopy is already proving useful
in a number of applications and showing promise in more.

2. Why does Tunneling Spectroscopy have Applications?

2.1 The Importance of Vibrational Spectroscopy

Tunneling spectroscopy is a sensitive technique for determining the vibrational
spectra of molecules. I want to emphasize, however, that in general we are
not interested in the vibrations themselves. We are usually interested in
using vibrational spectra as fingerprints to identify chemical species and
even to determine their orientation.

Non-chemists often do not fully appreciate the value of vibrational spectra
in the identification of chemical species. The simple fact is that vibrational
spectroscopy and NMR are the two most useful analytical techniques presently
available [1]. Because given functional groups (e.g., $C\equiv N$) always have vibrations in given ranges (e.g. 280 ± 10 meV for $C\equiv N$), one can tentatively
identify a compound by looking at correlation tables -- using reasoning like:
it has a vibration at 283 meV, therefore it may have a $C\equiv N$ group. Final identification is made by comparison to measured spectra of known compounds.

For bulk samples, infrared spectroscopy is almost always the most useful
technique for vibrational spectroscopy. Raman spectroscopy is used when

infrared is difficult or impossible (e.g., for seeing vibrations of molecules in solvents that are opaque to infrared or when the vibrations are not infrared active).

For adsorbed molecules on surfaces, infrared and Raman spectroscopy are difficult, primarily because of the small number of molecules that must be studied [2]. Since the interaction of an electron with molecular vibrations is much stronger than that of a photon, two electron spectroscopies are useful alternatives: electron energy loss spectroscopy and inelastic electron tunneling spectroscopy.

Before presenting a comparison of these various techniques, I would like to describe the capabilities of tunneling spectroscopy.

2.2 Spectral Range of Tunneling Spectroscopy

The useful range of most tunneling spectra has been from roughly 50 meV to beyond the range of molecular vibrations (beyond 500 meV). The recent development of a differential tunneling bridge [3] has extended the lower bound to 0 meV, thus including all molecular vibrations.

For example, Fig. 1 shows a tunneling spectra of benzoic acid liquid doped onto alumina. It was obtained by RICHARD KROEKER, who used a differential tunneling bridge and a magnetic field to drive the lead normal.

The differential tunneling bridge gives a difference spectrum between a doped junction and an undoped control junction on the same substrate. Thus, structure due to the metal electrodes and insulating layer is minimized. A magnetic field to drive the lead normal is necessary to make the lead phonon structure negligible.

The low energy vibrations now accessible to tunneling spectroscopy should contain a great deal of information -- including the fundamental vibrations of long chain molecules and the vibrations of entire adsorbed molecules relative to the substrate.

The spectral range has been extended beyond the range of molecular vibrations by the French group of J. KLEIN, A. LÉGER, and S. de CHEVEIGNÉ [4]. They have observed molecular electronic transitions at the higher energies. It may prove valuable in a number of applications, especially surface chemistry, to be able to obtain both vibrational and electronic level information.

2.3 Sensitivity of Tunneling Spectroscopy

There are many ways of measuring sensitivity. Three useful ones are: 1) minimum number of molecules that can be detected; 2) surface density of atomic species that can be detected; or 3) fraction of a monolayer of one compound that can be detected in the presence of another.

Figure 2 shows a small region of the tunneling spectrum for given percentages of p-deuterobenzoic acid included in a monolayer of benzoic acid liquid doped on alumina [5]. The C-D stretching vibration at 0.282 V can be resolved for 1%. This corresponds to: 1) 4×10^{10} molecules, 2) one deuterium atom per 1500 Å^2, and 3) 1% of a monolayer. This particular experiment was designed to optimize the surface density of atomic species that

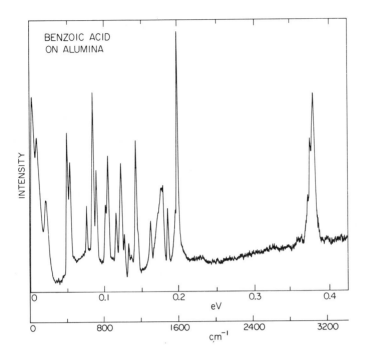

Fig. 1 Differential tunneling spectra of benzoic acid adsorbed on alumina. The doped junction had an area of 2 mm x 2 mm and a resistance of 75 Ω. The control junction had an area of 0.2 mm x 0.2 mm and a resistance of 25 Ω. Trace time was 4 hours with a 10 sec time constant. Voltage modulation was 0.7 mV rms at 2 K

could be detected (measure #2). Experiments designed to optimize the number of molecules would have smaller junction areas. We estimate that approximately one order of magnitude could be easily gained in this way.

2.4 Resolution of Tunneling Spectroscopy

The best resolution in published tunneling spectra is of order 1.1 meV, obtained with a modulation voltage of 0.6 mV rms at a temperature of 1 K. The majority of tunneling spectra, however, are taken at 4.2 K with resolutions between 2.1 and 4 meV, depending on the modulation voltage.

There is no known theoretical limit on resolution. The practical considerations that have limited resolution to date are:

1. The time to take a tunneling spectrum with a given signal to noise ratio varies more rapidly than $(1/\text{resolution})^4$.

2. It is more costly and time-consuming to transfer liquid helium into a research dewar and pump it down to of order 1 K than to insert a sample down the neck of a storage dewar at 4.2 K. It is still more costly and time-consuming to use He^3 or a dilution refrigerator to cool below 1 K.

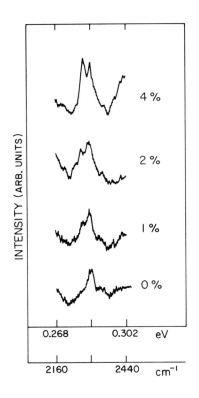

4 %

2 %

1 %

0 %

0.268   0.302   eV

2160   2440   cm⁻¹

Fig.2   Tunneling spectra of the
region around the 0.282 V C-D
stretch peak for various percentages
of p deuterobenzoic acid in a monolayer of benzoic acid liquid doped
onto alumina. The 1% curve repre-
the detection of 1 deuterium atom
per 1500 Å². The background peak
just above 0.282 V is believed to
be a combination band

2.5 Selection Rules of Tunneling Spectroscopy

Infrared active and Raman active vibrational modes both appear in tunneling
spectra with comparable magnitude [6]. There are, however, orientational
selection rules. The simplest orientational selection rule, that dipoles
oscillating perpendicular to the oxide couple more strongly to tunneling
electrons than dipoles oscillating parallel to the oxide, has been used by a
number of authors to deduce orientations of molecules on surfaces [7].

The real key to using tunneling peak magnitudes to obtain orientational
information will be the further development of tunneling theory.

2.6 Comparison of Tunneling Spectroscopy with Other Techniques

Infrared spectroscopy has certainly been the most widely used technique for
vibrational spectroscopy of adsorbed species [2]. Its advantages over tunneling spectroscopy are that:

1) It utilizes well developed technology.
2) It does not require cryogenic temperatures.
3) It can be applied to a wide variety of substrates.

Its disadvantages are:

16

1) Only limited spectral regions can be studied in general. Almost no work has been done in the very important range below 1000 cm^{-1}.
2) Its sensitivity is lower; it requires of order 10^3 more molecules to obtain a spectrum.

Raman spectroscopy has not been extensively used. The primary difficulty seems to be fluorescence by the substrate [8].

Electron energy loss spectroscopy is a relatively new technique [9]. In essence it involves measuring to high precision (7 meV is the current world record) the energy loss of low energy electrons (typically 5 eV) to the vibrations of adsorbed molecules. Its advantages are:

1) It can be done in-situ on metal surfaces that can be also studied by a variety of other surface spectroscopies (e.g., LEED and electron photo-emission).
2) There is no top metal electrode to shift vibrational modes.

Its disadvantages are:

1) It is an extremely difficult technical task to achieve an energy resolution an order of magnitude poorer than possible with tunneling spectroscopy. Only a few laboratories in the world have succeeded. Furthermore, prospects for improving this resolution appear small at present.
2) Charging effects make it difficult or impossible to apply to adsorption on insulating surfaces.

My own, speculative conclusions from these comparisons are:

1) Infrared spectroscopy will remain dominant in most cases where specific substrates (e.g. M-46 cracking catalyst [10]) must be studied.
2) Electron energy loss spectroscopy will grow in its applications. It will become dominant in the study of simple molecules on conducting substrates. Its lack of resolution will make it difficult to apply to molecules with tens of atoms.
3) Tunneling spectroscopy will be most useful for studying the adsorption of relatively complex molecules on model substrates: specifically, oxidized metals and supported metal particles.

3. What are the Applications of Tunneling Spectroscopy?

3.1 Surface Chemistry

In many ways tunneling spectroscopy is ideally suited to the study of chemical reactions on surfaces:

1) Monolayer or submonolayers of adsorbed molecules must be studied.
2) The chemical reactions that can confuse compound identification are precisely what is of interest.
3) The spectral range of tunneling spectroscopy allows examination of all vibrational modes, including those that correspond to the entire adsorbed molecule vibrating relative to the substrate.
4) Orientation information is important.
5) The resolution of tunneling spectroscopy is sufficient to study most molecules of interest.

Professor HENRY WEINBERG uses a wide variety of surface spectroscopies in his laboratory. He has also done pioneering work in the application of tun-

neling spectroscopy to surface chemistry [11]. Thus, he has an excellent
perspective on how tunneling spectroscopy fits in with other surface spectroscopies.

3.2 Identification of Biochemicals

The high sensitivity, spectral range, and relative lack of selection rules
make tunneling spectroscopy an intriguing possibility for the identification
of minute quantities of biochemicals. It could be expected that many of the
biochemicals would react with the aluminum oxide, but this is not a serious
disadvantage if the tunneling spectra are intended primarily as fingerprints.

The majority of work on the identification of biochemicals with tunneling
spectroscopy has been done by R. V. COLEMAN and his collaborators. He began
with a vacuum chamber that had a vacuum isolated doping chamber in which
compounds were heated and vapor doped onto the insulating layer.   More recently, he has used a liquid doping technique in which compounds are deposited
from solution onto the insulating layer. He has shown that very similar
nucleic acid derivatives can be differentiated with tunneling spectroscopy
[12]. This may have an important application to the sequencing of nucleic
acids.

3.3 Trace Substance Detection

A related application of tunneling spectroscopy is to the detection of substances present in trace quantities in either liquid solutions or in the air.
YELON and co-workers have shown, for example, that less than 10 ppm of
cyanoacetic acid in aqueous solution can be detected by simply dipping the
oxidized aluminum strips into the water [13]. Many researchers have made the
qualitative observation that junctions are sensitive to trace concentrations
of airborne molecules in the laboratory.

PROFESSOR YELON and co-workers have also shown that by dipping their
oxidized aluminum strips into solutions into which chemical reactions are
taking place they can sample reaction intermediates [14].

3.4 Adhesion

A great deal is known about cohesion. In fact, the majority of adhesive research is not concerned with how the adhesive adheres to the surface, but
how it sets (e.g., its polymerization). In part this reflects the real needs
of the industry. In part it reflects the lack of tools for studying the monolayer of adhesive molecules actually in contact with the substrate.

Tunneling spectroscopy may be useful in determining the basic mechanism
of adhesion for a number of adhesives.

3.5 Radiation Damage

Radiation damage to thin organic films is an area where several of the capabilities of tunneling spectroscopy are important. The sensitivity permits
measurement on small surface area samples that can be uniformly exposed with
moderate total beam intensities. The resolution provides detailed information
on the breaking of individual bonds within relatively complex molecules. The
spectral range may permit examination of disordering in polymers.

There have been a number of publications on electron beam damage since the original suggestions of M. PARIKH that tunneling spectroscopy might prove useful [15,16]. Since there will not be other talks in this conference on this application, I will reproduce a representative result here.

Figure 3 shows the tunneling spectra of five junctions that were fabricated close to each other on the same substrate by J. HALL [17]. They were all

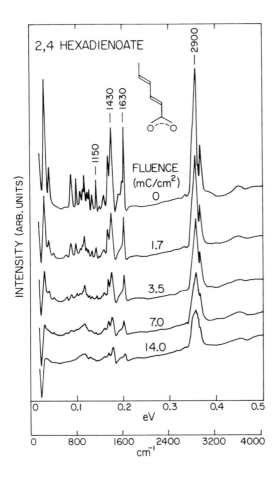

Fig.3 Tunneling spectra of five closely spaced junctions that were liquid doped with a solution of 2,4 hexadienoic acid in benzene. The spectra are labeled by the electron fluences that the junctions were exposed to in a scanning electron microscope. Since 1 mC/cm^2 corresponds to 0.625 electrons/Å^2, these spectra show that appreciable damage to a 2,4 hexadienote monolayer occurs before one electron/Å^2

liquid doped with the same drop of a solution of 2,4 hexadienoic acid in benzene. Four of the junctions (and thus the sandwiched molecules) were irradiated with differing fluences of 30 kV electrons in a scanning electron microscope. Damage cross sections for various bonds can be calculated based on the rate of decrease of their corresponding peaks in the spectra. For example, ν(C-H) peaks decreased at a rate corresponding to a damage cross section of 0.26 ± 0.04 (A^2/Incident 30 keV electron). For comparison, hexanoate ions had damage cross sections approximately five times larger, while benzoate ions had damage cross sections approximately seven times smaller. This can be qualitatively explained by the concept that molecules with more delocalized electrons are more resistant to damage.

Other experiments showed the stages of degradation of a simple biological molecule, fructose [18]. But, clearly, this work is only a start. Damage with other types of radiation (e.g., ultraviolet or proton beam) remains to be investigated. More sophisticated apparatus, specifically an ultra-high vacuum chamber with a helium temperature state and radiation source, would allow a researcher to measure successive tunneling spectra of the same junction during irradiation. It would allow damage experiments without the complications of a top metal electrode [16,19].

3.6 Lubrication

I believe that useful information about lubrication in general and lubrication of aluminum in particular will be found within the next few years with tunneling spectroscopy. The molecules that are typically used as additives are in the ideal size range of 10 to 30 atoms. The adsorption of these molecules from model lubricant solutions could be studied with tunneling sepctroscopy. These tunneling results could be correlated with macroscopic friction measurements on the same films.

[1] Work supported by the National Science Foundation

[2] Alfred P. Sloan Foundation Fellow (1975-1977)

References

1 See, for example, R.L. Shriver, R.C. Fuson, D.Y. Curtin: The Systematic Identification of Organic Compounds (John Wiley and Sons, New York, 1965)
2 L.H. Little: Infrared Spectra of Adsorbed Species (Academic Press, New York, 1966)
3 S. Colley, P. Hansma: manuscript in preparation
4 S. de Cheveigné, A. Léger, J. Klein: Proc. 14th Intl. Conf. Low Temp. Phys. Vol. 3, Otaniemi, Finland (North-Holland Pub. Co., 1975) p. 491
5 R. Kroeker, P.K. Hansma: to appear in Surface Science
6 M.G. Simonsen, R.V. Coleman, P.K. Hansma: J. Chem. Phys. 61, 3789 (1974); J. Kirtley, P.K. Hansma: to appear in Surface Science
7 See, for example, P.K. Hansma: 14th Intl. Conf. Low Temp. Phys. Otaniemi, Finland (North-Holland Pub. Co., 1975) Vol. 5, p. 264 (Figure 15).
8 T.A. Egerton, A.H. Hardin, Y. Kozirovski, N. Shepard: J. Catalysis 32, 343 (1974)
9 H. Ibach: J. Vac. Sci. Technol. 9, 713 (1972)
10 E.P. Parry: J. Catalysis 2, 371 (1963)
11 See, for example, B.F. Lewis, W.M. Bowser, J.L. Horn, Jr., T. Luu, W.H. Weinberg: J. Vac. Sci. Technol. 11, 262 (1974)

12 J.M. Clark, R.V. Coleman: Proc. Natl. Acad. Sci. $\underline{73}$, 1598 (1976)
13 Y. Skarlatos, R.C. Barker, G.L. Haller, A. Yelon: Surf. Sci. $\underline{43}$, 353 (1974)
14 Y. Skarlatos, R.C. Barker, G.L. Haller, A. Yelon: J. Phys. Chem. $\underline{79}$, 2587 (1975)
15 M. Parikh: 32nd Annual Meeting of Electron Microscope Society of America (Claitor Press, Baton Rouge, LA, 1974), p. 382
16 M. Parikh, J.T. Hall, P.K. Hansma: Phys. Rev. A $\underline{14}$, 1437 (1976)
17 J.T. Hall, P.K. Hansma, M. Parikh: to appear in Surface Science
18 P.K. Hansma, M. Parikh: Science $\underline{188}$, 1304 (1975)
19 J. Kirtley, P.K. Hansma: Phys. Rev. B $\underline{13}$, 2910 (1976)

II. Applications of Inelastic Electron Tunneling

Application of IETS to Surface Chemistry and Heterogeneous Catalysis [1]

W. Henry Weinberg [2]

Division of Chemistry and Chemical Engineering, California Institute of
Technology, Pasadena, CA 91125, USA

ABSTRACT

A brief review is given of Inelastic Electron Tunneling Spectroscopy (IETS)
as it is applied to investigations both of surface chemical phenomena as well
as heterogeneous catatlysis. The use of IETS to probe the vibrational structure of admolecules is compared with alternate electron scattering meth-
odologies, namely, Field Emission Energy Distribution (FEED) measurements
and Inelastic Electron Scattering Spectroscopy (IESS). Both the advantages
and the disadvantages of each type of measurement are emphasized.

1. Introduction

In order to understand fully the microscopic details of the conversion of
reactants to products over heterogeneous catalysts (i.e., in order to dis-
cover the elementary surface reactions), it is desirable to measure vibrational spectra. For example, the vibrational structural information allows
a molecular identification of reaction intermediates which in turn allows a
determination of the mechanism of the catalytic reaction. Although it is
possible to measure the vibrational structure of adspecies using optical
spectroscopy (e.g., absorption or reflection IR, and laser Raman spectroscopy),
the measurement of the energy loss of scattered electrons represents an alternate methodology which possesses a much greater sensitivity and thus is
inherently more appealing for the application of studying catalysis over small
area, reasonably well characterized catalytic surfaces.

Although, the thrust of this discussion is concerned with the inelastic
tunneling of electrons between two metal films through an oxide barrier in
a "planar" geometry or "diode" configuration (the IETS experiments), there
are two other types of inelastic electron scattering experiments which may
be performed and which lead to a vibrational energy loss spectrum. The first
type of experiment is the measurement of the energy distribution of electrons
which are field emitted in a "tip" geometry experiment in which a moderately
large potential applied to a field emitter in the shape of a very sharp
needle, the tip of which has a radius of curvature <1000Å, produces an elec-
trical field of sufficient strength to allow the tunneling of electrons
through the distorted potential barrier at the surface. Although still very
much in its infancy, preliminary results of vibrational structure observed
in these field emission energy distribution (FEED) measurements will be discussed below.

A second type of inelastic electron scattering experiment which resolves vibrational structure is the measurement of the kinetic energy distribution of a reflected electron beam after its interaction with a surface on which an adlayer is present, i.e., inelastic electron scattering spectroscopy (IESS) experiments, also known as characteristic electron energy loss measurements. This experiment is simple conceptually but is rather difficult technologically due to the requirement of a highly monoenergetic incident electron beam. This selection of energy (i.e., the use of an energy analyzer, and its attendant loss of intensity, in the production of the incident electron beam) is necessary due to the very small energy losses associated with the excitation of vibrational and/or phonon modes. Examples are given below of outstanding results of the utility of this experiment. The preponderance of the discussion is concerned, however, with inelastic electron tunneling spectroscopy (IETS) and the way in which it may be used in surface chemistry and heterogeneous catalysis research.

2. Field Emission Energy Distribution (FEED) Measurements

SWANSON and CROUSER [1] have observed perturbations of the FEED measurements due to the adsorption of phthalocyanine on both tungsten and molybdenum field emitters. These perturbations could be ascribed to three different mechanisms: (1) an elastic tunnel resonance resulting from a coupling of the conduction band of the metallic field emitter and electronic energy levels of the adsorbate lying within the conduction band of the metal; (2) inelastic scattering resulting in the excitation of low-lying electronic states of the adsorbate (electron-electron interactions); and (3) inelastic scattering resulting in the excitation of vibronic modes of the adsorbate (an electron-phonon type of interaction). Irreproducibilities were observed in the FEED measurements which were attributed to steric effects (the change in orientation of the adsorbate on the surface) and to partial decomposition of the adsorbate on the surface. Attempts to detect new structure in the FEED after adsorption of carbon monoxide were not successful, perhaps due either to the insensitivity of the electron energy analyzer or to the fact measurements could not be made at temperature below 77°K. In summary, using FEED measurements to detect vibronic excitations in admolecules is a promising, but not proved, surface spectroscopy. The effects of the high fields (e.g., in changing the orientation of the admolecules, or, indeed, field desorbing them) represents a challenge which must be met when carrying out this experiment.

3. Inelastic Electron Scattering Spectroscopy (IESS)

The experimental technique of IETS was pioneered by Professor F. M. PROPST at the University of Illinois [2,3]. He first demonstrated that IESS may be used to probe the vibronic structure of admolecules at a solid surface. In particular, vibrational excitations were observed in the case of H_2, N_2, H_2O and CO adsorption on a (100)W single crystal surface. Although the resolution of the electron energy analyzer used was only on the order of 50 meV, a great deal of information concerning the molecular nature of the various adsorbates could be ascertained, and various binding states of the admolecules on the surface could be resolved.

Professor H. IBACH and co-workers [4-9] have shown even more explicitly the utility of IETS by using an electron spectrometer with a higher resolution (on the order of 10 meV). Phonon loss peaks were observed for electron scattering from the (1Ī00) cleavage plane of ZnO [4,5] as well as the (111) surface of Si [4-6]. Of greater relevance to heterogeneous catalysis was the observation of vibrational modes due to the adsorption of O_2 on (111) Si as well as the contamination of the Si by the background gases in the ultrahigh vacuum chamber. A localized vibrational mode corresponding to a Si-H bond was observed after chemisorption of atomic hydrogen on a clean Si(111) surface [6].

More recently, IBACH and co-workers have observed the vibrational structure of oxygen and hydrogen on a W(100) surface [7,8] and of carbon monoxide on a Pt(111) surface [9]. In the latter, the CO was interpreted as being chemisorbed in an on-site position at low surface coverages and in both on-site and bridged sites at higher coverages. In similar studies, ANDERSSON has observed the vibrational structure both of oxygen [10] and carbon monoxide [11] chemisorbed on a Ni(100) surface. At low coverages, ANDERSSON also interprets his carbon monoxide measurements in terms of a linearly bonded complex to a single Ni atom. This interpretation is consistent with low-energy electron diffraction results, both experimental and theoretical [12]. In view of the fact the theory of IESS is rather elementary [13], it would appear this type of measurement represents a fertile tool insofar as surface chemistry and heterogeneous catalysis research is concerned. It was also shown that a total number of only 10^{11}-10^{12} oscillators is required for a measurable spectrum, and this is several orders of magnitude more sensitive than optical spectroscopies.

4. Inelastic Electron Tunneling Spectroscopy. (IETS)

The experimental probe of IETS represents a promising way by which surface phenomena important in catalysis, e.g., chemisorption and surface reactions, may be investigated. In IETS, a planar (diode) geometry is utilized consisting of a metal film, a thin (~ 25-30Å) oxide layer and a top metal film. When adsorbates are introduced onto the oxide surface prior to the evaporation of the top metal, the vibrational structure of the adspecies may be measured via the energy loss of electrons which tunnel inelastically through the barrier when a bias potential is applied across the two metal electrodes [14,15]. The second derivative of the tunneling current (d^2I/dV^2), easily measured electronically using a modulation and lock-in amplification detection scheme, is proportional to the oscillator strengths of the various vibrational modes of the molecular bonds which are detected. Thus, the preferred representation of the data is d^2I/dV^2 as a function of applied voltage since this is the IETS analogue of an IR spectrum. Typical IET spectra extend from ~30 to ~500 meV (i.e., from ~250 to 4000 cm^{-1}), although the upper limit can be extended to a few volts and is determined both by the breakdown voltage of the junction and the increased background present at the larger bias voltages. At voltages below ~30 meV, the phonon structure due to the metallic films dominates the IET spectra. Although numerous combinations of metal-oxide-metal junctions have been studied using IETS [15], emphasis will be placed on the Al-Al_2O_3-Pb junction for the following two reasons: (1) This junction is quite easy to prepare experimentally; and (2) The study of hydrocarbon reactions on alumina surfaces represents a technologically important problem. In this review, IETS, as it relates to surface chemistry and catalysis, will be discussed from the following points of view: (1) Experimental aspects; (2) The nature of the Al_2O_3 on which the IETS experiments are performed vis-à-vis 'real' alumina catalysts; (3) Theoretical concepts; (4) A brief review of the gas-surface interactions which have been studied using IETS; (5) A critical examination of the sensitivity, resolution and application of IETS; and (6) A preview of those areas in which IETS is likely to make a constructive impact.

The IETS effect was discovered by R. C. JAKLEVIC and J. LAMBE and discussed first in 1966 [16,17]. They not only discovered the effect, but, just as important, they gave a correct interpretation of the relevant physics. The impurity assisted tunneling spectra of water (hydroxyl groups), methanol, ethanol, acetic acid, propionic acid, cyanoacetic acid, and hydrocarbon pump oil present on the oxide surface in metal-oxide-metal tunnel junctions were measured for various combinations of junction materials. The majority of their work, however, was concerned with $Al-Al_2O_3-Pb$ junctions since oxidized Al forms a compact thin oxide rather easily, and a Pb overlayer is convenient due to its chemical inertness vis-à-vis molecular adsorbates on the oxide, its large ionic diameter which retards its diffusion into the oxide under the influence of an applied bias voltage, and its rather high superconducting transition temperature ($7.2°K$). It was also illustrated that IETS is sensitive to both IR and Raman (permanent and induced dipolar) active modes, that a representation of the data as the second derivative of the tunneling current with respect to bias voltage is the IETS analogue of optical IR and Raman spectroscopy, that the inelastic tunneling current corresponds to an increase in the conductivity of the junction (the first derivative of the tunneling current with respect to bias voltage) of approximately 1%, that the sensitivity of the technique is at least 1% of a monolayer (which represents approximately 10^{10} molecules per mm^2, and a sample the area of which is 1 mm^2 is easily useable), and that the IETS lineshape is understood in the sense that the spectra are broadened at higher measurement temperatures according to the Fermi-Dirac distribution function of the electrons in the metal films. In addition, it was shown that the resolution is further improved by having one or both of the metal films superconducting thereby creating a more sharply peaked electronic density of states function above the superconducting gap as compared to a normal metal.

Typical tunnel junction fabrication technique involves evaporation and planma discharge oxidation. The first step in the fabrication of an $Al-Al_2O_3-Pb$ junction is the evaporation of approximately 1000Å of Al in a bell jar the base pressure of which is $<10^{-6}$ torr. Then the Al is oxidized in a plasma discharge (approximately $50\mu_o$ of O_2 or H_2O) to form a barrier the thickness of which is on the order of 25Å. After in situ chemisorption or reaction studies, the final step in the junction preparation is the evaporation of Pb overlayer cross-strips of several thousand Ångstroms in thickness. Such junctions may be heated resistively and their temperature measured via the known (measured) temperature coefficient of resistance of the Al film [18].

Measurements in IETS invariably involve modulation and lock-in detection techniques. In the author's laboratory, the IET spectra are measured with a computer controlled analog detection system. Typically, the current through the sample is modulated with a 50 kHz sine wave. A lock-in amplifier detects the resulting 100 kHz voltage across the sample which is proportional to the second derivative of the current-voltage characteristic function of the junction. The steady state voltage is generated by a ramp generator which is controlled by the computer. A complete spectrum is obtained when the second derivative has been measured at several hundred different steady state voltages. The computer controlled analog detection scheme has three advantages over the more traditional analog schemes, namely, (1) Repeated fast scans and channel-by-channel sums reduce base line drift (low frequency noise); (2) Automatic multiplexing is possible (selected measurements of various windows of the spectrum); and (3) The data are in digital form immediately, facilitating data analysis via background subtraction, peak area integrations and functional deconvolutions.

Now that the way in which tunnel junctions are prepared and the way in which IETS measurements are made have been discussed, it is necessary to clarify the detailed nature of the barrier in the junction on which chemisorption and chemical reactions occur. By comparing IET spectra of barriers prepared in O_2 plasma at room temperature, and both an H_2O and a D_2O plasma at various temperatures from 300°K to 525°K, BOWSER and WEINBERG have shown that the barrier in IETS when prepared in an O_2 plasma is an oxide rather than an hydroxide of aluminum [19]. In this study, all junctions were of the Al-barrier-Pb type. It was further shown that by using low substrate temperatures in conjunction with H_2O or D_2O plasmas, hydrates of aluminum may be prepared, but the dehydration reaction is complete after heating the barrier to 525°K. Thus, barriers formed by an O_2 plasma at room temperature are identical to those formed by an H_2O plasma at 525°K. Incidentally, the ability to heat the oxide surfaces, as demonstrated for the first time in these experiments by BOWSER and WEINBERG, is of obvious and crucial importance in the application of IETS both to surface chemistry and heterogeneous catalysis [18,19].

Earlier, GEIGER et al. [20] had suggested that the barrier in Al-barrier-metal tunnel junctions may be an aluminum hydroxide rather than an aluminum oxide, but recent results of HANSMA et al. [21] are in agreement with those of BOWSER amd WEINBERG, namely, the barrier is in fact an aluminum oxide with OH groups present at the surface of the oxide, i.e., at the oxide-Pb interface in Al-oxide-Pb tunnel junctions. HANSMA et al. [21] have also made the very important comparison of catalytic activity of air oxidized aluminum versus the high surface area transition γ-alumina, of which the latter is a common industrial catalyst and catalyst support. It was found that the activity for 1-butene isomerization is actually somewhat greater for the air oxidized aluminum compared to γ-alumina after normalizing to specific surface area. Consequently, the film surfaces in tunnel junctions will be very representative models of high surface area transition aluminas used extensively on a commercial scale since air oxidized and plasma oxidized aluminum films are identical insofar as IETS measurements are concerned. In this same connection, KLEIN et al. [22] have investigated the MgO barrier in a Mg-MgO-Pb tunnel junction and have demonstrated that it is indeed an oxide as opposed, for example, to a hydroxide.

Prior to an examination of the kinds of surface chemistry results which have been reported using IETS, it is necessary to consider at least briefly the theory of IETS. Although a number of theories of IETS have appeared in the literature [23-29], the lack of a detailed understanding of the electron-phonon coupling constants represents a severe problem [14-30], and it is certainly true that a completely satisfactory theory of IETS is yet to be formulated.

Originally, SCALAPINO and MARCUS [23] applied the WKB approximation of tunneling through a barrier to IETS assuming charge-dipolar coupling between the tunneling electron and the adsorbate on the oxide barrier surface. The enhancement in the electron tunneling current due to this inelastic scattering was calculated using a Golden Rule formulation, and order-of-magnitude calculations for an Al-Al$_2$O$_3$-Pb junction with OH groups present on the Al$_2$O$_3$ were found to be in qualitative agreement with experimental measurements.

More recently, KIRTLEY et al. [28] have used a transfer Hamiltonian formalism to describe the IETS intensity of vibrational modes in organic molecules chemisorbed on the oxide barrier in metal-oxide-metal tunnel junctions. The theory made correct qualitative predictions of the magnitude of integrated intensities in IETS, the magnitude of Raman active versus IR active modes, the observability of optically forbidden modes, and the orientation of the admolecules on the oxide surface. This theory has the advantage of being more realistic than the original formulation [23], and, unlike many other previous theories [24-27], it may be used to determine the dependence of peak intensity on adsorbate concentration. Herein lies a difficulty, however. It is assumed that the effect of one admolecule on the tunneling electron is independent of the other admolecules. This leads inevitably to the conclusion that the peak intensities vary linearly with adsorbate concentration. In the one experiment which has tested this hypothesis, performed using tritiated benzoic acid (C$_6$H$_5$COOH) as the adsorbate in an Al-Al$_2$O$_5$-Pb junction, the intensity rather was found to vary as $n^{4/3}$ for ranges of adsorbate concentration n between 5% and 90% of saturation coverage [31].

Very recently, CUNNINGHAM et al. [29] have used a Golden Rule-transfer Hamiltonian formalism to calculate this inelastic tunneling probability which is proportional to the intensity measured in IETS. The potential which causes the inelastic transition is assumed to be due to a layer of dipoles at the adsorbate layer (as well as the infinite number of image dipoles which appear in both metal electrodes). When the cooperative effects of the adsorbate molecules are included, it was found that the model predictions agree quantitatively with the experimental results of LANGAN and HANSMA [31] insofar as the dependence of IET spectral intensity as a function of surface concentration is concerned. The theory of CUNNINGHAM et al. [29] should prove extremely useful to the surface chemist interested either in chemisorption or heterogeneous catalysis since the relation between IET spectral intensity and adsorbate concentration is of fundamental importance in this case.

Even prior to the development of the theory of CUNNINGHAM et al. [29], there have been a number of studies of gas-surface interactions using IETS. These have been concerned largely with chemisorption, but occasionally surface reactions have been investigated also.

LEWIS et al. [32] have shown clearly the way in which IETS may be used to determine the nature of various chemisorbed species. In particular, it was shown that H$_2$O adsorbs on Al$_2$O$_3$ dissociatively as hydroxyl groups, whereas formic acid and acetic acid chemisorb as formate and acetate ions, respectively. Moreover, the kinetics of adsorption of HCOOH were determined using IETS by measuring the surface coverage (via the intensity of the various vibronic excitations in the IET spectra) as a function of exposure of HCOOH to the Al$_2$O$_3$ surface. Finally, a coverage dependence of the energy at which a prominent C-H stretching mode is present in the IET spectra was observed. This could be attributed either to direct interadsorbate interactions, to indirect through-lattice interactions or to the effect of the top metal electrode [33]. These results represent one of the better examples of how IETS may be used to probe chemisorption, and by extension, heterogeneous catalysis (e.g., the unimolecular decomposition of the H$_2$O, HCOOH and CH$_3$COOH on the Al$_2$O$_3$ surface).

Subsequently, WEINBERG and co-workers [34,35] have investigated the mechanism of chemisorption of four different aromatic alcohols - phenol, catechol, resorcinol and hydroquinone - on an Al_2O_3 surface in an $Al-Al_2O_3$-Pb tunneling junction. In all cases, the chemisorbed species were found to be adions, and extensive hydrogen bonding was observed among the adions and between the adions and the OH groups present on the oxide surface. Moreover, relative peak intensities in the IETS measurements suggested that the phenol adsorbs with the aromatic ring very nearly perpendicular to the oxide surface, whereas the hydroquinone adsorbs with its ring much more nearly parallel to the oxide surface. This result is quite consistent with the structural properties of the alcohol molecules.

WALMSLEY et al. [36] have used IETS to monitor the chemical reaction between benzoyl chloride and the Al_2O_3 surface in an $Al-Al_2O_3$-Pb tunnel junction. The chemical reaction observed is, in fact, nothing more than the dissociative chemisorption of the benzoyl chloride in which the C-Cl bond is broken and the $C_6H_5CO^+$ combines with an oxygen from the Al_2O_3 lattice to form a surface benzoate species. The Cl^- subsequently combined with an H^+ from a surface OH group forming HCl which was pumped from the surface. An observed diminution of the OH stretching mode tended to substantiate this interpretation of the IETS results. Furthermore, it is interesting to note that a benzoate chemisorbed species is formed also on the Al_2O_3 surfaces, as judged by IETS, when either benzoic acid [31] or benzaldehyde [37] is chemisorbed.

In a fundamentally different type of chemisorption experiment SIMONSEN et al. [37] have demonstrated the feasibility of using IETS in conjunction with adsorption from the liquid phase onto the Al_2O_3 surface in an $Al-Al_2O_3$-Pb tunnel junction. This is of importance in the study of various macromolecules which would decompose prior to vaporization in the more usual gas phase adsorption experimental arrangement. It was also demonstrated explicitly that both IR and Raman active modes in anthracene are observed in IETS. Finally, the oxidation of benzaldehyde to aluminum benzoate by the Al_2O_3 surface was observed. An exciting prospect raised by this work is the possibility of studying liquid phase heterogeneously catalyzed reactions using IETS, but this concept is yet to be demonstrated experimentally.

Another recent and exciting application of IETS lies in the simulation of a supported metal catalyst. HANSMA et al. [38] have made IETS measurements of CO chemisorbed on Rh which is evaporated at various statistical coverages between 0.5Å and 4Å on the Al_2O_3 surface, i.e., they considered an $Al-Al_2O_3$-Rh-CO-Pb tunneling junction. They found a single bonding mode manifest by single Rh-CO and C-O stretching modes in the case of very low Rh coverages, and three Rh-CO modes and two C-O stretching modes at higher Rh coverages. These results were interpreted as linearly chemisorbed complexes on Rh monomers at low Rh coverages and bridged complexes on Rh clusters at higher Rh coverages. Unfortunately, less than saturation coverages of CO could not be investigated due to the rather poor vacuum conditions, but these results have opened up a new and extremely important application of IETS in view of the great commerical importance of supported metal catalysts.

To summarize, IETS has been discussed from several differeent points of view while keeping firmly in mind its applications to surface chemistry and heterogeneous catalysis. The way in which tunnel junctions are fabricated has been described, and measurement techniques have been discussed. The nature of the oxide barrier, especially in $Al-Al_2O_3$-Pb tunnel junctions, has been documented, and the important theories of IETS have been mentioned briefly. Finally, the most important applications of IETS to surface chemistry and heterogeneous catalysis have been cited. To conclude, a brief summary of the sensitivity, the resolution, and the applications of IETS to surface chemistry is given in Table 1.

Table 1 Summary of capabilities of inelastic electron tunneling spectroscopy

Resolution	~30 μeV (~0.25 cm⁻¹) if temperature and modulation functions are deconvoluted. ~1meV (~8cm⁻¹) otherwise.
Sensitivity	~5 x 10⁻³ monolayer (~2 x 10¹⁰ molecules)
Typical applications to surface chemistry and catalysis	Elucidation of the following: 1. Properties of oxide surfaces and metal-oxide interfaces. 2. Molecular structure of adsorbate and/or catalytic reaction intermediates or products of reaction. 3. Kinetics of adsorption and/or catalytic reactions. 4. Nature of interadsorbate and adsorbate-substrate interactions. 5. Orientation of adsorbate with respect to substrate. 6. Activation and/or poisoning of catalytic oxide surfaces. 7. Simulation of oxide supported metallic catalysts.

5. Summary

There are advantages and disadvantages associated with each of the three kinds of electron scattering vibrational spectroscopies discussed in this review. For example, the FEED measurements require a substrate which is compatible with the field emission microscopy (FEM) experiment, i.e., refractory metals are the easiest substrates with which to deal. Moreover, the large field associated with FEM (>10⁷ volts/cm) severely distorts the adparticles and, indeed, it may induce field desorption from the surface. In the IETS experiments, one must always be concerned with the effect that the top metal electrode may exert upon the adsorbate in this matrix isolation configuration. Moreover, in IETS, one does not have a real time record of the progress of the surface process (e.g., adsorption, surface reactions, etc.). Rather, one must be content with a discrete (in time) picture of the surface phenomena. The analogy is that a series of still photographs are obtained rather than a continuous motion picture, and the fixing of the photographs occurs upon evaporation of the top metal film. The principal advantages of IETS are its extreme sensitivity relative to most other vibronic spectroscopies and its high resolution (compared with IESS) due to the intrinsic energy selection of the electron beam if the tunneling measurement is performed at a low temperature, on the order of 4°K. The resolution is improved still further if one (or both) of the metal films is superconducting. The chief disadvantage to IESS is its rather poor resolution (~15 meV) relative to IETS (considerably better than 1 meV), but IESS is quite valuable since measurements may be made in real time of chemisorption and/or surface reactions on well characterized macroscopic (planar) surfaces.

Each of these experimental techniques have their characteristic advantages and disadvantages in measuring the vibrational structure of adsorbates and/or surface reaction intermediates. Each of them will be developed further, and each will take its place as an important adjunct to optical IR spectroscopy in the study of the vibrational structure of admolecules. Their great advantage is that they are very suitable to studies on well characterized surfaces of low area.

[1]Work supported by the National Science Foundation under Grant Number GK-43433

[2]Camille and Henry Dreyfus Foundation Teacher-Scholar and Alfred P. Sloan Foundation Fellow.

References

1   L.W. Swanson, L.C. Crouser: Surface Sci. 23, 1 (1970)
2   F.M. Propst, T.C. Piper: J. Vac. Sci. Technol. 4, 53 (1967)
3   J.F. Wendelken, F.M. Propst: Rev. Sci. Instrum. 47, 1069 (1976)
4   H. Ibach: J. Vac. Sci. Technol. 9, 713 (1972)
5   H. Froitzheim, H. Ibach: Z. Physik 269, 17 (1974)
6   H. Froitzheim, H. Ibach, S. Lehwald: Phys. Letters 55A, 247 (1975)
7   H. Froitzheim, H. Ibach, S. Lehwald: Phys. Rev. B14, 1362 (1976)
8   H. Froitzheim, H. Ibach, S. Lehwald: Phys. Rev. Letters 36, 1549 (1976)
9   H. Froitzheim, H. Hopster, H. Ibach, S. Lehwald: to be published
10 S. Andersson: Solid State Commun. 20, 229 (1976)
11 S. Andersson: Solid State Commun. 21, 75 (1977)
12 S. Andersson, J.B. Pendry: to be published
13 D.L. Mills: Surface Sci. 48, 59 (1975)
14 C.B. Duke: Tunneling in Solids, Academic Press, N.Y., 1969
15 E. Burstein, S. Lundqvist, Eds.: Tunneling Phenomena in Solids, Plenum Press, N.Y., 1969
16 R.C. Jaklevic, J. Lambe: Phys. Rev. Letters 17, 1139 (1966)
17 J. Lambe, R.C. Jaklevic: Phys. Rev. 165, 821 (1968)
18 W.M. Bowser, W.H. Weinberg: Rev. Sci. Instrum. 47, 583 (1976)
19 W.M. Bowser, W.H. Weinberg: Surface Sci. 65, 000 (1977)
20 A.L. Geiger, B.S. Chandrasekhar, J.G. Adler: Phys. Rev. 188, 1130 (1969)
21 P.K. Hansma, D.A. Hickson, J.A. Schwarz: J. Catal., to be published
22 J. Klein, A. Léger, M. Belin, D. Défourneau, M.J.L. Sangster: Phys. Rev. B7, 2336 (1973)
23 D.J. Scalapino, S.M. Marcus: Phys. Rev. Letters 18, 459 (1967)
24 L.C. Davis, Phys. Rev. B2, 1714 (1970)
25 C. Caroli, R. Combescot, D. Lederer, P. Nozieres, D. Saint-James: J. Phys. C 4, 2598 (1971)
26 G.K. Birkner, W. Schattke: Z. Physik 256, 185 (1972)
27 T.E. Feuchtwang: Phys. Rev. B 10, 4135 (1974)

28 J.R. Kirtley, D.J. Scalapino, P.K. Hansma: Phys. Rev. B14, 3177 (1976)
29 S.L. Cunningham, W.H. Weinberg, J.R. Hardy: to be published; see also these Proceedings
30 N.O. Lipari, C.B. Duke, R. Bozio, A. Girlando, C. Pecile, A. Padva: Chem. Phys. Letters 44, 236 (1976)
31 J.D. Langan, P.K. Hansma: Surface Sci. 52, 211 (1975)
32 B.F. Lewis, M. Mosesman: W.H. Weinberg, Surface Sci. 41, 142 (1974)
33 J.R. Kirtley, P.K. Hansman: Phys. Rev. B12, 531 (1975)
34 B.F. Lewis, W.M. Bowser, J.L. Horn, Jr., T. Luu, W.H. Weinberg: J. Vac. Sci. Technol. 11, 262 (1974)
35 W.H. Weinberg, W.M. Bowser, B.F. Lewis: Jap. J. Appl. Phys., Suppl. 2, Pt. 2, p. 863 (1974)
36 D.G. Walmsley, I.W.N. McMorris, N.M.D. Brown: Solid State Commun. 16, 663 (1975)
37 M.G. Simonsen, R.V. Coleman, P.K. Hansma: J. Chem. Phys. 61, 3789 (1974)
38 P.K. Hansma, W.C. Kaska, R.M. Laine: J. Am. Chem. Soc. 98, 6064 (1976)

IETS with Applications to Biology and Surface Physics [1]

R.V. Coleman, James M. Clark, and C.S. Korman

Department of Physics, University of Virginia
Charlottesville, VA 22901, USA

ABSTRACT

Applications of inelastic electron tunneling spectroscopy to the study of
biological molecules will be reviewed. Experiments include work on amino
acids, purine and pyridine bases, nucleosides, nucleotides, DNA, RNA, proteins and other selected compounds. Detailed studies of mononucleotides
using a series of derivatives with selected bases have established that the
tunneling spectra resolve small differences in chemical structure as well as
clearly distinguishing different bases. The technique has also been extended
to polynucleotides and may be of help in sequence studies. Applications include studies on ultraviolet radiation damage and a series of results on
bases, mononucleotides, and polynucleotides will be prepared. Results and
problems associated with applying the tunneling technique to the full DNA
molecule and other large molecules will also be analyzed. Applications to
other relevant materials such as long chain fatty acids will also be discussed

1. Introduction

Vibrational spectra of a wide range of organic molecules have been studied
using inelastic electron tunneling spectroscopy [1-4] (IETS). The resulting
resolution and intensity give the technique a significant potential for
spectroscopic work, and both experiment and theory show that Raman and infrared types of interactions can contribute to the tunneling mode intensity.
Initial experiments on small biological molecules such as amino acids and
long chain fatty acids showed good spectra which correlated well with infrared and Raman work [2]. Subsequent extension to large biological molecules
[2-5] such as DNA and RNA showed characteristic spectra, but resolution of
individual modes was difficult due to low intensity and overlapping modes
although similar difficulties are encountered in other spectroscopies. CLARK
and COLEMAN [6] carried out extensive work on nucleoside and nucleotide derivatives and showed that sufficient resolution could be obtained with IETS
in order to differentiate among a variety of nucleotide derivatives.

The development of the technique involves the adsorption of the molecule
on an oxide substrate, and this process plays an important role in determining
the intensity and resolution which can be obtained for a particular molecule.
In this paper we describe a range of spectra obtained for biological molecules
and attempt to analyze some of the factors which influence the quality and
resolution of the spectra obtained from IETS. Some of these factors are
chemical reactions of side groups with the substrate, effects due to orientation of the molecule with respect to the substrate and relative intensity
contributions from Raman type (polarizability) or infrared type (dipole) interactions [7,8] of the tunneling electrons with specific vibrational modes
of the molecule. In order to demonstrate some of these factors and to assess
their role in the IETS of the larger biological molecules, we will also include results on a number of smaller organic ring compounds where rather
detailed analysis of the surface interaction can be made.

2. Amino Acids

IETS spectra of amino acids were initially obtained by SIMONSEN and COLEMAN [9] using a vapor doping technique. The majority of these spectra were on simple amino acids such as serine, cysteine and lysine containing mainly CH, NH, CO and OH groups. The spectra were dominated by modes attributed to CH bonds with some additional structure attributed to bonds such as C=O, C-NH$_2$, and CH$_2$-OH. Subsequent work by SIMONSEN, COLEMAN and HANSMA [2] using liquid doping gave spectra of much better resolution and a more complete comparison to infrared spectra could be accomplished. For a small amino acid such as glycine good agreement with infrared results was obtained and the mode assignments based on infrared work were quite precise as shown in Fig. 1. Spectra of the more complex amino acids were also obtained and a spectrum obtained from L-phenylalanine is shown in the upper curve of Fig. 2. The resolution of individual modes was quite good and identifications based on infrared work are indicated in Fig. 2. However, the identifications involving

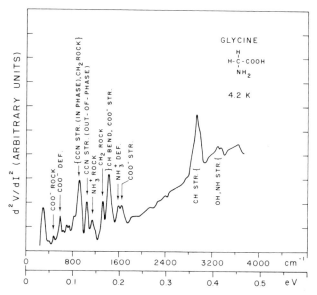

Fig.1 Tunneling spectrum of glycine. Junction doped with H$_2$O solution.

vibrations of the side group should be viewed as tentative since the modes of the aromatic ring are contributing substantial intensity to the IETS spectrum. For example, the IETS spectrum of acetophenone as shown in the lower curve of Fig. 2 is very similar with respect to frequency and relative intensity. In this case nearly all modes are associated with the ring and have been assigned using IR and Raman results as listed in Table V of section 7. The role of the ring relative to the side group would appear to be similar in the two cases and suggests that the ring makes major contributions to the intense modes which are observed in both spectra. The similar modes are probably characteristic of a single aromatic ring attached to a carbon atom of an adsorbed side group. The specific adsorption of the side group to the alumina substrate appears to play a significant role and may be required in order to obtain an intense IETS spectrum. This point will be discussed more fully in later sections.

35

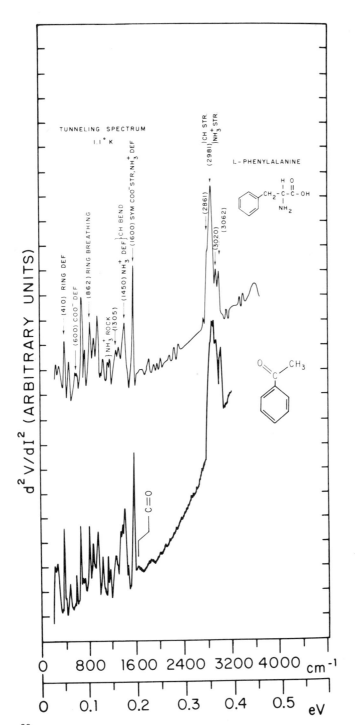

Fig.2 Tunneling spectra of L-phenylalanine (200 Ω) (H$_2$O solution) (upper curve) and acetophenone (800 Ω)(alcohol solution) (lower curve). The two spectra show a strong similarity in both mode energy and relative intensity

3. Nucleoside and Nucleotide Derivatives

Extensive IETS spectra on the nucleotide units of DNA and RNA have been obtained by CLARK and COLEMAN [6]. These molecules consist of a sugar, a
pyrimidine or purine base, and a phosphate group. They form the backbone
structure of DNA and RNA as shown in Fig. 3. The IETS spectra give a large
number of well-resolved modes and reasonably good agreement with the Raman
spectra of these same compounds in solution has been obtained. Shifts in
frequency for particular modes can be attributed to the pH dependence of the
solution spectra. Examples of IETS spectra obtained for the four nucleotide
units comprising the DNA backbone are shown in Fig. 4(a) and 4(b).

The strongest modes below 1600 cm^{-1} in these spectra can generally be
assigned to the normal modes of the base ring. Figure 5 shows an amplified
tunneling spectrum of adenosine-5'-monophosphate in the range below 1600 cm^{-1}.
The majority of the strong modes between 500 and 1600 cm^{-1} are characteristic

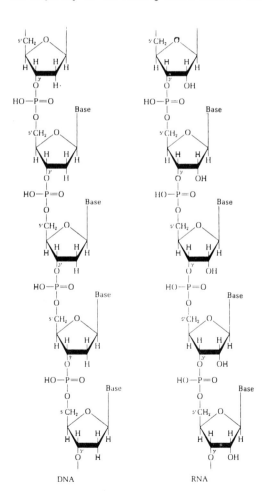

Fig.3 Backbone structure
of DNA (deoxy-D-ribo-
nucleic acid) and RNA (D-
ribonucleic acid). The
four bases in DNA are
adenine (A), guanine (G),
cytosine (C) and Thymine
(T), and the four bases in
RNA are adenine, guanine,
cytosine and uracil (U)

37

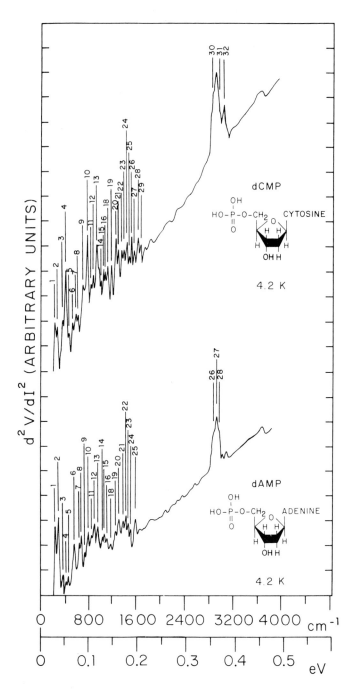

Fig.4(a) Tunneling spectra of deoxycytidine-5'-monophosphate (dCMP) (2545 Ω) and deoxyadenosine-5'-monophosphate (dAMP) (1032 Ω). Both from H_2O solutions

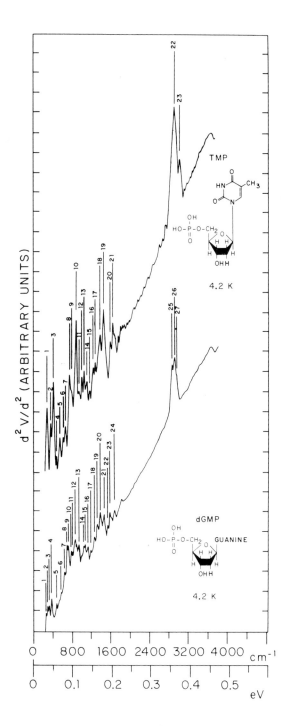

Fig.4(b) Tunneling spectra of thymidine-5'-monophosphate (TMP) (1679 Ω) and deoxy-guanosine-5'-mono-phosphate (dGMP) (1063 Ω). Both from H₂O solutions

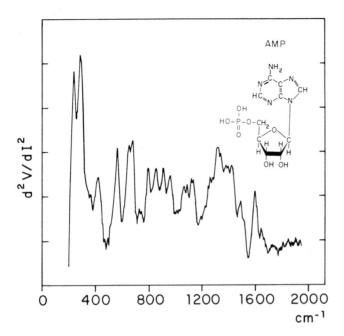

Fig.5 Amplified tunneling spectrum of AMP (750 Ω) (H_2O solution) in the range 200 - 2000 cm^{-1}. Strongest modes are contributed by the adenine ring

of the adenine ring and have been shown to recur in all IETS spectra of adenine derivatives [6]. This result is consistent with Raman results where the bases A, G, C, and U, by virtue of their aromaticity, are expected to generate intense Raman scattering. Raman lines due to vibrations of ribose or deoxyribose are ordinarily weak except in the CH stretching region (2800-3000 cm^{-1}). The tunneling spectrum of D-ribose alone is dominated by strong CH modes, but shows relatively weak modes at other wavenumbers [6]. Vibrational modes involving the phosphate group are expected near 980 and 1090 cm^{-1}. They do not appear to make dominant contributions to the IETS spectra since modes in this region are present in both the nucleoside and nucleotide spectra. These general observations make it quite easy to unambiguously identify the base associated with different nucleotide spectra. For example Table I lists the wavenumbers for the modes observed in the IETS spectra of 5'-dAMP and 5'-dCMP as shown in Fig. 4(a). The wavenumbers observed for the strong modes associated with the adenine base in 5'-dAMP are underlined in the left-hand column of Table I. The spectrum of 5'-dCMP does not show strong modes at these wavenumbers while the IETS spectra measured for eight different adenine derivatives [6] consistently show these strong adenine modes. Different derivatives containing the same base can however be differentiated by comparing the modes below 600 cm^{-1}. IETS modes in the range 200-550 cm^{-1} are sensitive to small variations in the molecular structure such as the attachment site of the phosphate on the sugar, the type of sugar (D-ribose or D-deoxyribose) and the presence or absence of the phosphate groups. CLARK and COLEMAN [6] have listed the detailed wavenumbers and have shown expanded IETS spectra below 600 cm^{-1} which clearly demonstrate this point.

40

Table I   Wave numbers measured by IETS for dAMP-5' and dCMP-5'

	dAMP-5' cm^{-1}		dCMP-5' cm^{-1}
1.	234	1.	246
2.	294	2.	278
3.	367	3.	375
4.	428	4.	428
5.	464	5.	480
6.	557	6.	544
7.	645	7.	593
8.	678	8.	625
9.	742	9.	726
10.	799	10.	786
11.	871	11.	843
12.	907	12.	883
13.	956	13.	948
		14.	984
14.	1028	15.	1016
15.	1069	16.	1057
16.	1113	17.	1085
17.	1178	18.	1125
18.	1230	19.	1194
19.	1250	20.	1270
20.	1306	21.	1315
21.	1371	22.	1363
		23.	1399
22.	1420	24.	1448
23.	1472	25.	1492
24.	1512	26.	1532
25.	1585	27.	1569
		28.	1637
		29.	1686
26.	2883	30.	2883
		31.	2935
27.	2924	32.	2996
28.	2964	33.	3061

The molecular configuration of the nucleosides and nucleotides would generally maintain the plane of the sugar and the plane of the base at right angles. Whether this relative orientation is maintained on the aluminum oxide substrate is not known at present. The possible role of this relative orientation on tunneling intensity is not precisely known either. These points will be discussed in more detail below.

4. IETS Intensity Enhancement with COOH Substituents

The IETS spectra of benzoic acid and benzoic acid derivatives have been studied extensively and analysis has shown that they adsorb on the aluminum oxide substrate in a benzoate configuration due to the loss of the proton from the COOH side group and the formation of a CO_2^- ion bonded to the aluminum atoms. Benzoic acid and derivatives always show very strong IETS intensity which is connected in some way with the interaction of the COOH group with the alumina substrate. The addition of a COOH group as a substituent to a series of ring compounds containing from one to five rings consistently produces a strong increase in IETS intensity as compared to the IETS spectrum of the same ring structure without the COOH substituent. Examples are shown in Figs. 6(a) and 6(b) for anthroic acid and ferrocene carboxylic acid. Both anthracene and ferrocene alone show a relatively low IETS intensity which is not appreciably changed for a wide range of junction resistance and modulation level.

This large change in intensity observed for the major ring modes as a function of ring substituent has previously been observed for small ring compounds as reported by KORMAN and COLEMAN [10]. A change in ring orientation relative to the substrate induced by the change in substituent could possibly explain this result through a dependence of the tunneling intensity on orientational selection rules [7]. KORMAN and COLEMAN [10] examined this possibility in terms of mode assignments for the various ring modes. Although mode assignments consistent with the expected orientational selection rules are possible, critical choices for a few strong modes make the analysis uncertain. Changes in ligands and electronic charge distributions with change of substituent could also play a role and the relative contributions of all these factors to tunneling intensity are not yet sorted out.

Although the intensity variations are not completely understood it is clearly desirable to develop methods for enhancing the tunneling intensity. The addition of a COOH group is one way to do this and we have explored this possibility relative to enhancing the IETS spectra of nucleotides. The experiments were carried out with the base uracil and nucleotide derivatives containing uracil. As a first step the tunneling spectra of uracil and orotic acid (6-carboxyuracil) were compared as shown in the upper two curves of Fig. 7. A clear intensity enhancement is observed with the addition of a COOH group as well as the appearance of a strong C=O mode at 1682 cm^{-1} where only a weak one is observed in uracil. This would be consistent with a ring orientation which would more favorably orient the free C=O bond. A protonation of the C=O due to interaction with the OH sites on the alumina to form C-OH could also account for the loss of C=O intensity in uracil. The substitution of the COOH group on the ring could then inhibit the protonation and restore C=O intensity. A very strong additional mode at 512 cm^{-1} is also observed.

The substitution of orotic acid as the base in orotidine-5'-monophosphate has almost no effect on the tunneling intensity of the nucleotide (3rd curve from top Fig. 7) although a fairly strong ring mode at 434 cm^{-1} associated

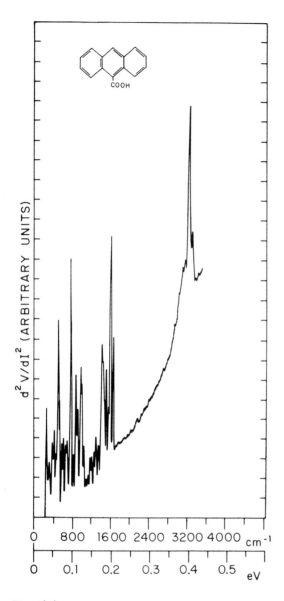

Fig.6(a) Tunneling spectrum of anthroic acid (500 Ω) (alcohol solution)

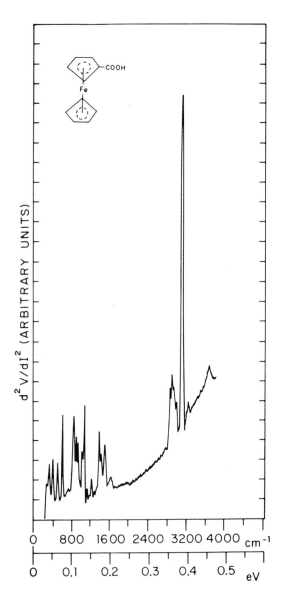

Fig.6(b) Tunneling spectrum of ferrocene carboxylic acid (530 Ω) (alcohol solution)

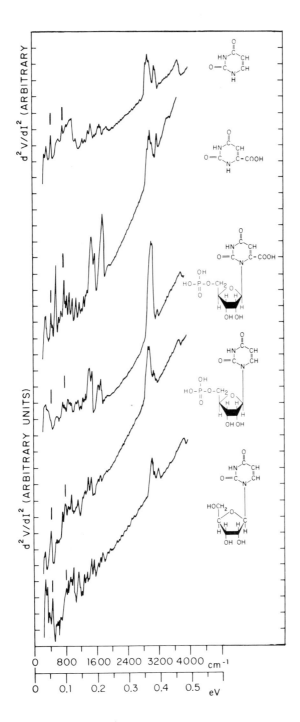

Fig.7 From top to bottom tunneling spectra of uracil (100 Ω), orotic acid (130 Ω), orotidine-5'-monophosphate (200 Ω), uridine-5'-monophosphate (314 Ω) and uridine (269 Ω). All doped from H_2O solution

with uracil and also present in orotic acid is absent from the OMP spectrum. The lower two curves of Fig. 7 show the tunneling spectra obtained on the nucleotide UMP and the nucleoside uridine. The presence or absence of the phosphate group has practically no effect on the spectrum while the substitution of orotic acid for uracil produces a shift in relative intensity but no major enhancement of intensity. A comparison of the wavenumbers for the modes in the two nucleotides and the nucleoside shows the majority of the modes to occur at approximately the same wavenumber suggesting that all three curves are essentially characteristic of the UMP tunneling spectrum. Table II gives

Table II Wavenumbers for the Raman and IETS spectra of UMP and the IETS spectrum of OMP

(IETS)OMP Peak Pos. cm^{-1}	(IETS)UMP(disodium) Peak Pos. cm^{-1}	(Raman)UMP pH 1.6	7.4	12.6
307	280			
	357			
379				
	434			
561	551	562	558	
	587			593
617	623	630		
			640	
722	724			
762				
799	796	786	784	790
855		820	810	
	869			
		885	880	875
907	909			
960	954			
	986		982	980
1000		1000	1000	
	1022			1020
1057	1055			
		1090		
1113	1119		1125	
1173				
1242	1240	1234	1232	1242
		1260		
1290	1301			1296
	1373			
1387				1380
		1400	1397	
	1450			1460
1468		1475	1470	
	1579			
1617				
	1631			1640
1686	1684		1685	
1722				

a comparison of the measured IETS wavenumbers for OMP and UMP along with a comparison to the Raman spectra obtained from solution at three different pH values. A reasonable correlation between all three spectra can be seen.

The addition of the COOH group to uracil clearly produces the intensity enhancement effect observed in the other simple ring compounds, however, this enhancement does not appear to transfer to the nucleotide. Whether this indicates a change in surface binding, a change in relative molecular orientation, or a change in molecular bond structure for the more complex molecule is not clear. The intensity enhancement effect is cleary specific and further work will have to be done to see if substituent changes exist which can enhance the IETS intensity in the more complex biological molecules.

5. Radiation Damage to Nucleotides

The IETS technique has been used to make some preliminary studies of uv radiation damage to nucleotide derivatives. The radiation was carried out in an ultra high vacuum system (10^{-8} torr) and the junction was exposed to uv light of wavelengths in the range 2100-4000 Å. The molecules were exposed to uv radiation at the intermediate doping step before the deposition of the lead electrode.

Initial results for UMP are shown in Fig. 8 and show IETS spectra obtained from junctions irradiated for 0, 2, and 10 minutes. The strong ring modes associated with the uracil base and indicated by the vertical lines in Fig. 8 are rapidly reduced in intensity indicating that the initial primary damage occurs to the base ring. Further exposure then begins to reduce the CH mode structure as well. The experiments were carried out with the junctions mounted on a liquid nitrogen cooled block. Junction resistances before and after irradiation were comparable and in the range several hundred ohms indicating that the results are not associated with evaporation of the molecule from the substrate.

The preliminary results look promising but more detailed experiments at precise wavelengths and exposure times will have to be carried out in order to fully evaluate the technique.

6. Nucleic Acids and Proteins

As mentioned in the previous section intensity enhancement becomes an increasingly important problem in the more complex molecules. IETS spectra have been obtained for ribonucleic acids, deoxyribonucleic acids and a variety of proteins such as hemoglobin and myoglobin. The junctions can be made quite easily, but the resolution and identification of all of the individual modes becomes more difficult. We will give representative examples in this section with a selected comparison to Raman spectra and will also point out some interesting points concerning the intrinsic tunneling intensity and structure observed for these large biological molecules.

An IETS spectrum of calf thymus DNA is shown in Fig. 9 for the range 200 to 1800 cm^{-1}. Thirty-three modes have been measured in this range and Table III gives a listing along with a comparison to the Raman spectrum obtained for solid calf thymus DNA. The correspondence is extremely good considering the uncertainty introduced by the low intensity observed for many of the tunneling and Raman modes. The Raman spectrum for calf thymus DNA in H$_2$O solution does not fit the tunneling data as well as does the solid DNA spectrum. This is expected since the tunneling spectrum should be more characteristic

47

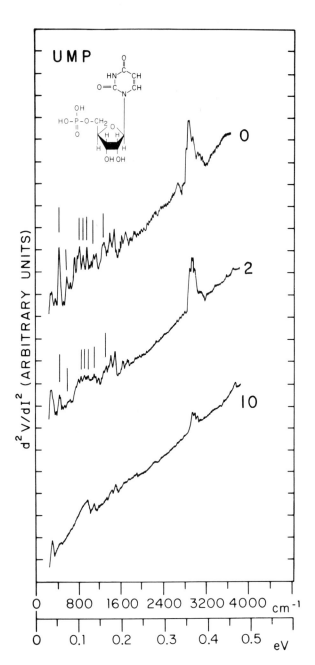

Fig.8 Tunneling spectra of UMP as a function of uv radiation exposure. From top to bottom (0 min. exposure, 370 Ω), (2 min. exposure, 1700 Ω), (10 min. exposure, 218 Ω). Doped from H_2O solution

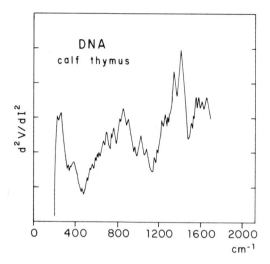

Fig. 9 Amplified tunneling
spectrum of calf thymus DNA
in the range 200-2000 cm^{-1}.
(2000 Ω from H$_2$O solution).
Modes and wavenumbers are
compared to Raman spectrum
in Table III

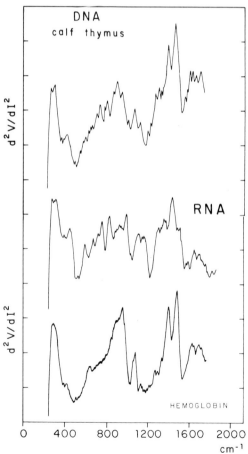

Fig.10 Amplified tunneling
spectrum of calf thymus DNA
(2000 Ω) yeast RNA (1113 Ω)
and hemoglobin (987 Ω) in
the range 200-2000 cm^{-1}

of the dry DNA, and the presence of water is known to shift some of the vibrational modes in DNA.

The major peaks and overall shape of the IETS spectrum of the large biological molecules appears similar in many cases and may be governed by the presence of a large amount of CH structure in the molecule. This observation is demonstrated in Fig. 10 where IETS spectra of DNA, RNA and hemoglobin are shown. The strongest modes between 1300 and 1600 cm^{-1} can be assigned to the CH asymmetric and symmetric deformation modes in all three cases. In the lower range below 1000 cm^{-1} the aluminum oxide phonon structure contributes to the similar overall envelope observed for all three curves. In the case of DNA and RNA many additional medium and weak peaks are observed which correlate

Table III   Wavenumbers for IETS and Raman spectra of calf thymus DNA

	IETS cm^{-1}	RAMAN solid cm^{-1}
1.	263	
2.	295	
3.	368	
4.	415 m	415 wbr
5.	494 w	496 s
6.	566 w	570 w
7.	592 w	597 w.
8.	621 w	625 w
9.	637 w	
10.	657 w	666 m
		683 w
11.	698 m	
12.	729 s	731 s
13.	776 sh	
14.	795 s	786 vs
15.	858 m	
16.	889 s	873
17.	940 m	
		963 w
18.	1000 m	1013 m
19.	1058 s	1062 m
20.	1113 m	
		1144 vw
21.	1189 m	1183 m
22.	1215 m	1208 m
23.	1258 m	1247 s
24.	1294 m	1306 s
25.	1315	
		1335 s
26.	1371 s	1372 s
27.	1447 s	1449 w
28.	1537 m	1537 w
29.	1557 m	
30.	1581 ms	1586 s
31.	1608 m	1612 w
32.	1639 m	
33.	1682 ms	1670 s

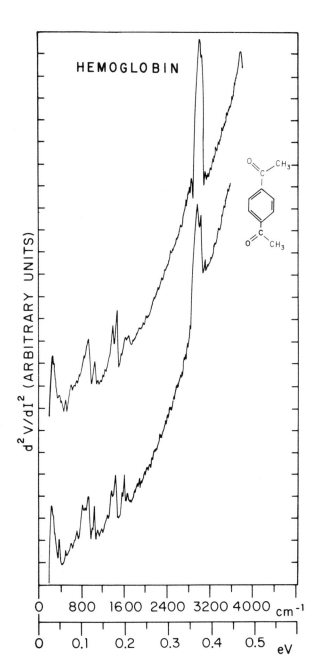

Fig.11 Tunneling spectra of hemoglobin (3520 Ω) (upper curve) (H$_2$O solution) and diacetylbenzene (1500 Ω)(lower curve) (alcohol solution). Both spectra show similar features and are dominated by CH mode intensity

well with modes contributed by the base residues of the molecule. In the case of hemoglobin as shown in the lowest curve of Fig. 10 very little fine structure can be resolved. It would appear that the detailed structure of the α and β chains contributes very little to individual tunneling mode intensity. This feature of the hemoglobin tunneling spectrum is reemphasized in Fig. 11 where the IETS spectra of hemoglobin and diacetylbenzene are compared. Both tunneling spectra show essentially the same major peaks and background shape leading to the conclusion that the main units of the hemoglobin are very weakly coupled to the tunneling electrons. The only relatively strong structure is contributed by CH groups and the Al-O phonon structure. In the case of diacetylbenzene the ring modes are essentially quenched, and again the most prominent modes are CH modes (bending, stretching and rocking). The reason for decoupling of the ring modes is not definitely established although the diacetylbenzene interacts with the alumina substrate. Both acetyl groups appear to bond with little evidence of any free C=O mode intensity suggesting that the ring is parallel to the surface. The junction resistance for hemoglobin was 3200 Ω and the junction resistance for diacetylbenzene was 1500 Ω indicating a fairly strong molecular coverage in both cases. Further evidence of variations in ring mode intensity for the acetophenone series will be presented in section 7.

The hemoglobin molecule has four heme groups with the basic structure arising from protoporphyrin IX. Two CH_2COOH groups are attached to the protoporphyrin IX ring structure and these may interact with the alumina substrate in a way similar to the interaction of the acetyl groups of diacetylbenzene. However, the IETS spectrum of protoporphyrin IX also gives very low intensity for the individual modes. An amplified IETS spectrum of protoporphyrin IX is shown in Fig. 12. A large number of weak modes can be de-

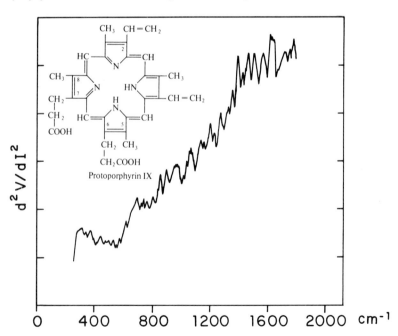

Fig.12 Amplified tunneling spectrum of protoporphyrin IX in the range 200-2000 cm^{-1}. (2540 Ω) (H$_2$O solution)

52

tected, although the limit of resolution is being pushed in this experiment. These results indicate that as the molecular units increase in complexity the IETS mode resolution becomes more of a problem. Further experiments with mode enhancement due to surface interactions of substituent groups may be helpful, but initial evidence indicates this to be less important in the larger molecules. Surface interactions can produce drastic changes in the IETS spectra observed for simple aromatic ring compounds and evidence of this will be discussed in the next section.

7. Effects of Surface Interactions on Tunneling Intensity

Wide variations in tunneling intensity for single ring compounds can be observed as changes are made in the substituent groups on the ring. The observations suggest that these intensity variations are correlated with the adsorption of the substituent side group on the alumina substrate. Examples of these intensity changes are shown in Fig. 13 for a series of acetophenone derivatives. The lowest curve in Fig. 13 shows the very intense IETS spectrum obtained for p-acetyl benzoic acid. The spectrum shows a very strong C=O mode at 1686 cm^{-1} and the overall spectrum is quite similar to other p-substituted benzoic acids adsorbed on the alumina in the benzoate configuration. The unmodified C=O bond suggests that the acetyl group does not interact with the surface in this case. The tunneling spectrum of acetophenone is again rather intense as shown in the next higher curve of Fig. 13, but no appreciable C=O mode is observed suggesting that the acetyl group is involved in the adsorption on the alumina surface. As pointed out in section 2 the spectrum appears to be characteristic of all single unsubstituted aromatic rings attached to the carbon atom of an adsorbed side group. Table V shows the comparison of the IETS spectrum to the assigned infrared and Raman modes for the free molecule of acetophenone. Very good correlation is obtained for most modes, although there are definite frequency shifts which may be associated with the surface adsorption of the acetyl group.

Acetophenone (methyl phenyl ketone) is a strong adsorber on the alumina substrate and junction resistance can be built up to large values. The spectrum in Fig. 13 was measured on a junction with a resistance of 800 Ω although excellent IETS intensity can be obtained over a range of resistance from 200-2000 Ω. The reactions of aldehydes and ketones are mainly those of nucleophilic attack on the carbon atom of the carbonyl group and those due to the enolizable hydrogen atoms α to carbonyl of the type

$$-CH_3 - \overset{\prime}{\underset{O}{C}} \rightleftarrows -CH_2 = \underset{OH}{C} -$$

Water molecules can be coordinated at the Lewis acid sites on the alumina to form Bronsted acid sites which could act to catalyze the reactions. Although such reactions are possible on the alumina surface it is more likely that the bonding occurs at the negatively charged oxygen of the carboxyl group. This can bond directly to the aluminum atom or through an intervening OH group of the Bronsted acid site. Variations in the tunneling intensity could then be due to changes in the detailed stereo-chemistry associated with the bonding.

As shown in the 3rd highest curve of Fig. 13 p-substitution of an OH group on acetophenone suggests that the surface bonding now involves the OH group leaving a free acetyl group which again gives rise to a strong C=O mode at 1665 cm^{-1}. The higher frequency ring modes are in precise agreement

53

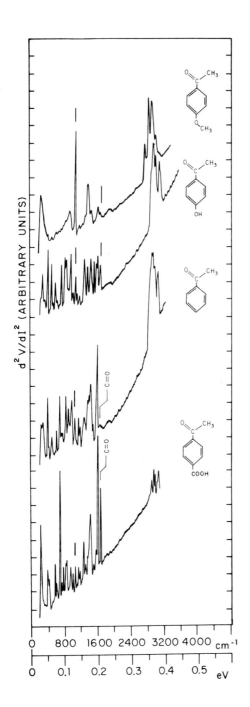

Fig.13 From top to bottom, tunneling spectrum of p-methoxyacetophenone (200 Ω), p-hydroxyacetophenone (350 Ω), acetophenone (800 Ω) and p-acetylbenzoic acid (800 Ω). All doped from alcohol solution

Table IV Comparison of Raman, IR and IETS wavenumbers for p-hydroxyaceto-phenone

IETS cm^{-1}	IR and RAMAN (free molecule)[†] cm^{-1}	Wilson No.	
290 s			
355 w			
428 vs			
	483 m	6 a	IP
508 s	500 m	16 b	OP
548 w			
	574 s	γCO	
601 m	592 m	γOH asy.	
645 w	636 w	6 b	IP
	697 w	4	OP
738 s	729 w	12	IP
	785 s	1	IP
823 s			
	840 vs	17 b	OP
859 s	850 sh	βCO	
964 s	970 s	νC-(CH$_3$)	
1016 m	1028 m	18 a	IP
1085 s	1087 m	7 a	IP
1117 m	1117 m	18 b	IP
1174 m	1173 vs	9 a	IP
1290 vs	1288 vs	13	IP
1363 s	1370 vs	14	IP
1436 vs	1444 vs	19 b	IP
1524 s	1518 s	19 a	IP
1573 s	1589 s	8 b	IP
1605 s	1606 m	8 a	IP
1665 s	1654 vs	νCO	

[†]OP out-of-plane; IP in-plane

with the assigned infrared and Raman modes of the free molecule as shown in Table V. However, the IETS modes below 600 cm^{-1} show considerable shifts due to perturbations connected with the surface adsorption and precise assign-ment of these modes is uncertain.

When a methoxy group is p-substituted in acetophenone a rather drastic change in the IETS spectrum is observed as shown in the highest curve of Fig. 13. The resulting spectrum indicates that the ring modes have been quenched in intensity while the major peak intensity is connected with the adsorbed side group. The spectrum is in fact nearly identical to the IETS spectrum observed by many investigators for the adsorption of HCOOH (formic acid) on aluminum oxide [3,4]. The peaks in the range 2800-3200 cm^{-1} are associated with symmetric and asymmetric C-H stretching modes with possibly some contributions from combination bands. The broad peak with a maximum at approximately 950 cm^{-1} (118 meV) is assocaited with the Al-O modes and

Table V   Comparison of Raman, IR and IETS wavenumbers of acetophenone

IETS cm^{-1}	IR and RAMAN (free molecule) cm^{-1}	Wilson No.
343 m		
	369	6 a
412 s	412	16 a
	468	16 b
514 m		
577 w	589	γCO
625 m	618	6 b
	693	4
702 s		
	731	1
768 w	761	11
851 s	848	βCO
924 ms	926	$\overset{\cdot}{17}$ b
	955	νC-CH$_3$
980 ms	974	5
1040 w	1025	18 a
1071 m	1076	18 b
1121 sh		
1165 m	1158	9 b
1194 w	1178	9 a
1269 w	1265	13
1294 mw	1302	3
1325 w	1330	14
	1358	δ_sCH$_3$
1385 m		
	1433	δ_{as}^{+}CH$_3$
1452 ms	1451	19 b
1502 s	1490	19 a
1605 vs	1599	8 a
1674 m	1684	νCO

has a characteristic tail extending to lower energies. This broad peak is always prominent in tunneling spectra when it is not dominated by specific strong modes from the dopant compound.

The extremely intense mode at 1064 cm^{-1} (133.4 meV) is also observed as an intense mode in the IETS spectrum of formic acid and could be identified with an out-of-plane C-H deformation vibration. An alternative interpretation is a combination mode of the reacted group. For formic acid adsorbed on alumina LEWIS, et al. [3] observed this same peak at 131.4 meV and assigned it to the C-H deformation vibration. Peaks are also observed at 147, 162 and 172 meV in the tunneling spectrum of p-methoxyacetophenone while LEWIS, et al. [3] observed peaks at 147, 162, and 172.6 meV for the formic acid tunneling spectrum and assigned them to in-plane deformation vibrations of the C-H bond. The broad peak at ~200 meV can be assigned to the C=O and C-O in-plane stretching vibrations although the C=O would appear to be greatly reduced in intensity as compared to the free acetyl group.

From their analysis LEWIS, et al. [3] concluded that HCOOH chemisorbs on the Al_2O_3 substrate at room temperature as a formate ion in the configuration shown below

$$Al - O -C\overset{\displaystyle \nearrow O}{\underset{\displaystyle \searrow H}{}} \; .$$

The reacted acetyl group and the formate ion both involve C=O and C-H bonds and may be very similar in structure. The reason for the loss of the ring mode structure in p-methoxyacetophenone is not clearly established. It may be due to ring orientation or it may be due to changes in the chemisorption and bond structure.

The characteristic tunneling spectrum observed for p-methoxyacetophenone is also observed for other p-substituted acetophenones as shown for p-cyanoacetophenone in the upper curve of Fig. 14. The lower curve in Fig. 14 shows the IETS spectrum of p-substituted nitroacetophenone where strong intensity is restored to the ring modes and a strong C=O mode is also observed. In this case the NO_2 group is bonded to the alumina surface. The general conclusion is that the detailed adsorption involving the acetyl group and the resulting tunneling spectrum is determined by whether the p-substitution involves a non-bonding or bonding substituent relative to the alumina substrate. In the case of a surface non-bonding p-substituent the characteristic IETS spectrum is unaffected by the relative electron donating or withdrawing properties of the substituent.

8. Conclusions

Inelastic electron tunneling spectroscopy can be applied to the entire range of biological compounds and in many cases a well-resolved spectrum comparable to Raman and infrared results can be obtained. This has been demonstrated for acmino acids and for a series of nucleotide units of DNA and RNA. For molecules of large molecular weight and complex structure the resolution of individual modes becomes a more difficult problem and can be quite variable depending on the structure of the particular molecule.

The results on nucleotides have shown that the strongest intensity in the tunneling spectrum arises from the base residue and that identification of the base from IETS can easily be accomplished. For nucleotide derivatives

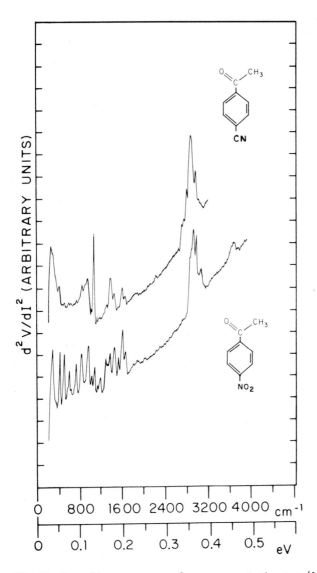

Fig.14 Tunneling spectrum of p-cyanoacetophenone (140 Ω) and p-nitro-acetophenone (650 Ω). Doped from alcohol solution

with the same base but with modifications of the sugar or point of attachment of the phosphate group the low energy IETS modes below 600 cm^{-1} can be used for identification of the different derivatives. The wavenumbers observed for these low lying modes are very sensitive to small changes in molecular structure and studies on a series of adenine derivatives have demonstrated this point.

Interactions with the aluminum oxide substrate can also shift the low energy modes and this has been demonstrated for COOH substitution in uracil and uridine-5'-monophosphate. In the case of 6-carboxyuracil the interaction of the COOH with the substrate also enhances the IETS intensity. This intensity enhancement effect is not directly transferred to the nucleotide either due to a change in the surface interaction or a change in molecular orientation on the surface. The possibility of IETS intensity enhancement through selective substitution needs further study since it can substantially improve IETS resolution. For small aromatic ring compounds this intensity enhancement has been consistently demonstrated for compounds containing from one to five rings. Whether similar effects can be observed in non-aromatic structures and more complex molecules is not known.

The importance of the surface interaction at the alumina substrate has also been demonstrated for the IETS spectra obtained on a series of acetophenone derivatives. The IETS spectrum is highly sensitive to whether the p-substituted group in acetophenone is surface bonding or non-bonding. In the case of a bonding group like COOH or OH the spectrum shows strong ring mode intensity and evidence of a free acetyl group. In the case of a surface non-bonding p-substituent the acetyl group bonds and the ring modes are quenched with the strong IETS modes arising only from the bonded acetyl group configuration.

The results in this paper have shown that IETS has a wide range of application to vibrational mode analysis of molecules in combination with modifications due to surface interactions. Further development of the technique will be needed before a complete assessment of its advantages and disadvantages can be made.

Acknowledgements

The authors wish to thank Professor J. E. Coleman, Professor G. H. Coleman, Professor O. R. Rodig, Professor T. H. Cromartie, and Professor F. S. Richardson for valuable discussions regarding the chemistry of the compounds studied in this paper. The authors wish to thank Professor J. W. Mitchell for suggesting the acetophenone series and supplying some of the chemicals.

[1] Work supported by the U.S. Energy Research and Development Administration Contract E-(40-1)-3105 and National Science Foundation Grant GB3481.

References

1 J. Lambe, R.C. Jaklevic: Phys. Rev. 165, 821 (1968)
2 Michael G. Simonsen, R.V. Coleman, Paul K. Hansma: Journ. of Chem. Phys. 61, 3789 (1974): Michael G. Simonsen, R.V. Coleman: Phys. Rev. B8, 5875 (1973)
3 B.F. Lewis, M. Mosesman, W.H. Weinberg: Surf. Sci. 41, 142 (1974)
4 J. Klein, A. Leger, M. Belin, D. DeFourneau, M.J.L. Sangster: Phys. Rev. B7, 2336 (1973)
5 Paul K. Hansma, R.V. Coleman: Science 184, 1369 (1974)
6 James M. Clark, R.V. Coleman: Proceedings of the National Academy of Sciences 73, 1598 (1976)

7   John Kirtley, D.J. Scalapino, P.K. Hansma: Phys. Rev. B14, 3177 (1976)
8   R.C. Jaklevic, J. Lambe in "Tunneling Phenomena in Solids", edited by
    E. Burstein and S. Lundqvist (Plenum, New York, 1969), Chap. 18, p. 243
9   Michael G. Simonsen, R.V. Coleman: Nature 244, 218 (1973)
10 C.S. Korman, R.V. Coleman: Phys. Rev. B15, 1877 (1977)

Application of IETS to Trace Substance Detection [1]

Arthur Yelon

Departement De Genie Physique, Ecole Polytechnique
Montreal, Canada H3C 3A7

ABSTRACT

As IETS is very sensitive compared to IR or Raman spectroscopy, it is very promising for trace substance detection, especially for aqueous solutions. We review the work which has been done to date on IETS of solution doped junctions, and show that both qualitative and quantitative measurements should be possible with this technique. Some of the steps which must be taken to make the method generally interesting are reviewed. Finally, the application of IETS to a practical problem, the ozonation of phenol in water, is described.

It has been known for over a decade, since the first work of JAKLEVIC and LAMBE [1], and of SCALAPINO and MARCUS [2], that IETS is very sensitive, yielding the possibility of detecting a few molecules per cm^2. This extreme sensitivity is illustrated by the difficulty in eliminating the C-H stretch near 360 meV, and the virtual impossibility of eliminating the OH stretch at 450 meV from samples oxidized in "pure" O_2 [3]. It was also shown in the early work that tunneling is almost equally sensitive for IR and Raman active modes [1,2]. These facts make IETS extremely interesting as a technique for the detection of trace substances.

In order for the technique to work, three things are needed in addition to the high sensitivity: the possibility of detecting the trace substance in the presence of large quantities of other substances, the possibility of identifying the substance, and the possibility of determining its concentration. We shall deal with each of these problems in the course of this paper. Obviously, one can imagine investigating traces in either vapor or liquid phase environments. As all of the earliest IETS work involved adsorption from the vapor phase, it was natural that the first trace substance work should be done in a similar way. By 1970, KLEIN and LEGER [4], had detected a large number of pollutants in Paris air by this method.

While vapor phase doping clearly works, and has recently been applied to the study of tobacco smoke [5], this does not seem to be an area of great promise. Both IR and Raman spectroscopy can be used in air; and Raman, in particular has been used for remote sensing of air pollution [6]. It would be quite inconvenient in contrast to have to expose an electrode at the mouth of a chimney in order to perform IETS on air pollution. Clearly the greatest advantage of performing trace substance detection by IETS should be realized with solutions, especially aqueous solutions since water strongly absorbs in the IR.

The first results on solutions doping were reported in 1973 and 1974 by three independent groups: BOGATINA et al. [7], SKARLATOS and co-workers [8], and HANSMA and co-workers [9,10]. The experimental techniques and analytical methods used by the various groups to perform solution doping will be discussed in detail below. The basis of the method is to expose the first electrode (which may already be oxidized or not, depending upon the reactivity of the chemicals involved) directly to some of the solution, after which the solvent is evaporated.

In Fig. 1, we show tunneling spectra obtained with two junctions whose first electrodes were dipped in pure and in 10 ppm acetic acid respectively.

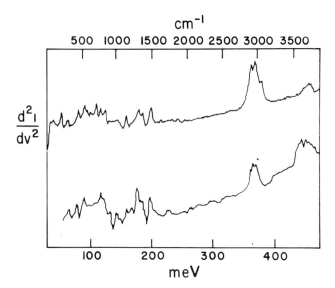

Fig.1 Electron tunneling spectra of junctions exposed to pure (top) and 10 ppm (bottom) CH_3COOH, showing changes in intensities and in peak locations. (After SKARLATOS et al. [8])

In. Fig. 2, we show a tunneling spectrum obtained by dipping in concentrated formic acid, and a spectrum obtained with formic acid adsorbed from the vapor phase (from LEWIS et al. [11]). From these two figures, we can immediately answer two of the questions we have raised above. First, while there are changes between the spectra for 10 ppm and concentrated CH_3COOH, these are relatively minor compared to the similarities. In particular the C-O, C=O, and C-H peaks differ by only 1 to 2 meV. The presence of large quantities of water does not mask the acetic acid spectrum. Second, the spectrum for HCOOH from solution [8] is essentially the same as that from the vapor phase [11,12], differing usually by less than 2 meV, and at most by 5 meV. Further, the HCOOH and the CH_3COOH spectra are sufficiently different that they cannot be mistaken one for the other. Thus qualitative analysis and identification of traces are possible with this technique.

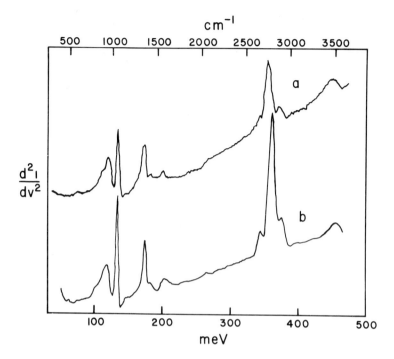

Fig.2 Electron tunneling spectra of junctions exposed to HCOOH. (a) Solution doped with concentrated liquid. (After SKARLATOS et al. [8]). (b) Vapor doped (After LEWIS et al. [11])

To date, in addition to the two organic acids just mentioned, a significant number of organic substances have been detected in solution doped junctions. These include phenol [13], benzoic acid [14,15], stearic acid [16], and glyoxilic acid [13] all from aqueous solution. The last of these, glyoxilic acid, seems to be the only trace substance detected in liquid phase doped junctions which was not deliberately introduced into the junction. We shall return to this, below.

Anthracene has been introduced into tunneling junctions from chloroform and from THF [10], TCNQ and its derivatives have been introduced from water, acetone, and acetonitrile solutions [15]. Considerable effort has been expended upon solution doping with biological substances. These include adenine [7,10], L-phenylalaline [9,10], uracil [9], DNA [9,10], t-RNA [9,10], glycine [10], hemoglobin [10], as well as biological molecules with substituent groups [10]. The solvents used have included water, alcohol, benzene, and chloroform. In general, there seems to be no reason why any of the inorganic, organic, and biological materials which have been used in vapor phase doping [17] cannot be introduced by solution doping, frequently with better results. In addition some substances which are difficult to introduce from the vapor may be introduced from solution.

Obviously, for IETS to attain its ultimate utility in trace analysis as in other domains, it will be necessary to establish standards and even atlases of spectra, as has been done for IR. One will need to take some precautions when surface reactions are possible [10,18] but this should not represent too great an obstacle.

Most of the experiments listed above have been performed with concentrated solutions rather than with trace concentrations. But it is clear that IETS should provide enough sensitivity for many applications. In Fig. 3, we show the results of SKARLATOS et al. [8] for the C-H peak of cyanoacetic acid, near 360 meV, and the C≡N peak at 280 meV. It is clear that the C-H at least should be detectable well below ppm concentrations, and that C≡N should be detectable to at least 10 ppm concentrations. In fact, with reactive chemicals, it may be desirable to use dilute solutions, in order to avoid producing junctions which are too thick and too unstable [13].

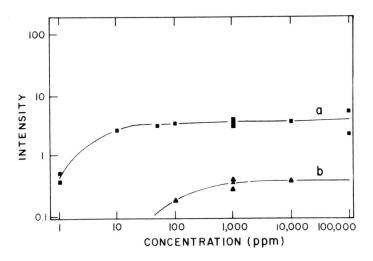

Fig.3 Plot of intensity versus solution concentration for $CNCH_2COOH$ doped junctions. (a) C-H peak at 360 meV. (b) C≡N peak at 280 meV. (After SKARLATOS et al. [8])

The problem of establishing quantitative analysis of trace substances by IETS is considerably more difficult than that of establishing qualitative analysis. From Fig. 3, we see that while the intensity of the d^2I/dV^2 tunneling signal is a monotonically increasing function of solution concentration, it is exceedingly nonlinear. This is due primarily to a nonlinear relationship between solution concentration and concentration of adsorbed species on the surface. This is presumably due to the competition between the trace molecules and the solvent molecules for adsorption sites on the electrode or insulator surface. If the trace substance is a large organic molecule and the solvent is water, it is not surprising to find that most of the sites will be occupied by water.

Recently LANGAN and HANSMA [14] have measured the concentration of adsorbed species on the surface directly by a radioactive tracer technique. They have compared the dependence of absolute surface concentration upon solution concentration with that for signal intensity. Their results are shown in Fig. 4. We see that the two dependences, while similar are certainly not the same. Before penetrating further into this subject, let us

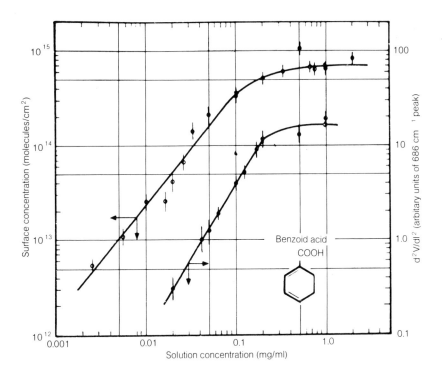

Fig.4 Surface concentration (left scale) and IETS peak intenstiy (right scale) a junction of solution doped with C_6H_5COOH. (After LANGAN and HANSMA [14])

return to the question of the experimental techniques and analytical methods used by the various groups which have performed solution doping experiments to date.

Almost all of the solution doped junctions studied thus far have consisted of Al-oxide-Pb structures, which combine ease of fabrication with the signal enhancement obtainable when the counter electrode contains a large atom [19]. However, BOGATINA and co-workers used Pb-oxide-Pb without difficulty, and one can even imagine using synthetic barriers [20], although this would tend to lead to complications in the spectra. All of the junctions were fabricated at room temperature, except those of BOGATINA et al. [7] who held the first electrode at nitrogen temperature, to keep the organic molecules from desorbing before the second electrode was produced. The solution has been brought into contact with the first electrode simply by dipping into the solution [8,13,16], by pipeting a drop into its surface [7,9,10,14,15] and by introducing the solution into the vacuum system under controlled atmosphere [13]. All of these methods give comparable results. Most of the experiments on solution doped junctions have been performed on samples which had simply been allowed to dry, but HANSMA, COLEMAN and their co-workers [9,10, 14,15] have spun-dry their samples in order to obtain more reproducible junctions. This is very difficult to achieve in an absolute way, as the impedance of the junction and the measuring circuitry used will affect the absolute heights of inelastic tunneling peaks. The impedance is a function of junction history, such as time at room temperature, voltages applied, the times for which they were applied etc. Theoretically [1,2,21], the intensity of an IETS peak, is believed to be proportional to the concentration of that bond in the junction. But if one can not reproduce and maintain junction impedance and measurement conditions, how is one to make use of this fact?

If one is interested in the relative intensities of different functional groups in the same molecule [22] this problem is clearly not too serious. If one is interested in comparing the orientations of similar molecules [15] it may be sufficient to produce junctions of roughly comparable impedance, and to measure them on the same apparatus. But if one wishes to compare concentrations for many junctions, over a period of time, as one would normally do for quantitative analysis, this is not sufficient. Is there any way of obtaining a measurement which is directly related to concentration of a given functional group in the junction? LANGAN and HANSMA [14] seem to suggest that there is not, but we believe that there is a very simple solution to this problem.

SKARLATOS and co-workers [13] suggested that the number of OH groups contributing to the stretch peak at 450 meV is essentially the same in "pure" junctions as it is in doped junctions. They concluded this on the basis of comparison of "pure" and water doped junctions of roughly equal impedance. If this assumption, that the junction is always saturated with OH, is correct, then a more realistic comparison between the peaks in two different junctions is obtained by comparing peaks normalized with respect to the 450 meV peak, rather than by comparing absolute intensities. The curves of Fig. 3 were obtained in this way. This has two immediate advantageous consequences: first, the relative intensity of a peak remains constant with time, even if the impedance of the junction changes; second, the relative peak heights obtained for two samples from the same solution were essentially the same, which would not have been the case for absolute intensities. These observations suggest that the initial hypothesis was indeed correct.

If one applies the normalization procedure just described to the data of LANGAN and HANSMA [14] it appears from the two experimental curves given in their paper, that this would yield a curve parallel to that of absolute concentration. As this approach is far easier to use than an elaborate experimental procedure to obtain reproducible results, it seems to be the most appropriate way to approach quantitative measurement. It would also seem to be useful as an approach to the comparative surface chemistry of similar molecules [15]. However, we must note that it is still necessary, in all cases, to establish empirically the relationship between solution concentration and surface concentration. This may involve considerable work in some cases for example if two different organic molecules are competing for the same adsorption sites. Finally, we must note that our assumption that there is a linear relationship between surface concentration and intensity seems to be contradicted by the latest theoretical work of CUNNINGHAM et al. [23] who find that tunneling current should depend upon $n^{4/3}$. How this would affect our conclusions is not entirely clear.

If IETS is to be a serious tool for trace substance detection, it must be possible to detect substances which have not been deliberately introduced into the junction. In this sense, there has been only one "practical" experiment of this kind thus far. This is the work of SKARLATOS et al.[13] on the ozonation of phenol in water. Phenol may be present as an impurity in drinking water, giving it an objectionable taste and odor [24]. The most effective way of eliminating it is ozonation. As the ozonation proceeds, the solution becomes amber colored (or yellow for very dilute solution). After all the phenol has been ozonated, the solution becomes colorless again.

The objectives of the study were to detect phenol in the ppm range, to detect its disappearance, and to attempt to identify the compounds responsible for the yellow color of the solution. Two models [25,26] had been given for the sequence of ozonation reactions, but it had not been demonstrated which of these was correct. In Fig. 5, we show spectra for a 100 ppm solution,

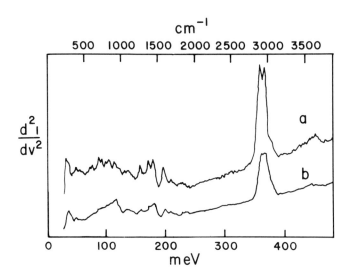

Fig.5 Electron tunneling spectra of junctions exposed to 100 ppm C_6H_5OH solution. (a) Original solution. (b) After ozonation for 5 minutes. (After SKARLATOS et al. [13])

and for the same solution after ozonation for 5 minutes. It is clear that the phenol has been eliminated in the second case. By comparison with Fig. 1 we can see that the end product is acetic acid, which is known to be generated in the process. At this concentration, the yellow color was never observed.

In order to observe the intermediate products, 1% solutions were ozonated, until they were deeply colored. It was impossible to produce useful junctions from these solutions, as these were noisy when fabricated and became too highly resistive in a short time. By doping from a diluted solution, however,

it became possible to produce usable junctions. In Fig. 6, we show the
spectrum of such a junction. In addition to the peaks due to phenol, as in
Fig. 5, there are extra peaks, indicated by the small arrows. By comparison
with its IR spectrum, it was possible to identify the product responsible
for these peaks as glyoxilic acid. As this is expected from one of the proposed series of reactions [25] but not from the other [26], it was possible
to conclude that one model is certainly incorrect and that the other is probably correct. As this answered a long standing question of serious concern
to water chemists, it seemed to us to be a convincing demonstration of the
power of IETS in trace substance detection.

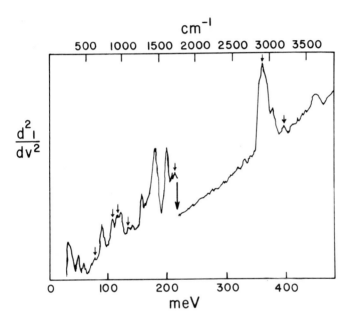

Fig.6 Electron tunneling spectrum of partially oxidized C_6H_5OH solution,
showing the phenol peaks as well as new structure, indicated by small arrows,
glyoxilic acid, and attributed to HCOCOOH. The long arrow indicates a shift
of the zero. (After SKARLATOS et al. [13])

There are many problems of the sort we have just described to be faced in
solution chemistry and especially in water chemistry, for which IR or Raman
spectroscopy are far from being the best tools. It is to be hoped that
chemists will accept the challenge, and adapt IETS for these problems. Much
work needs to be done before the technique can achieve its full utility, but
we believe that the effort will be rewarded.

The author wishes to thank Y. SKARLATOS for helpful discussions.

[1]Supported in part by the National Research Council of Canada.

References

1 R.C. Jaklevic, J. Lambe: Phys. Rev. Lett. 17, 1139 (1966); J. Lambe,
   R.C. Jaklevic: Phys. Rev. 165, 821 (1968)
2 D.J. Scalapino, S.M. Marcus: Phys. Rev. Lett. 18, 459 (1967)
3 Reviewed in R.G. Keil, T.P. Graham, K.P. Roenker: Appl. Spectrosc. 30,
   1 (1976)
4 J. Klein, A. Leger: private communication
5 J. Walachova: in Prac. Konf. Cezk. Fyz. Pr., 3rd, 1973, M. Matyas, Ed.
   (Academia, Prague, 1974), p. 154
6 R.L. Beyer: Proc. IEEE 59, 1644 (1971); Opt. Quant. Elect. 7, 147 (1975)
7 N.I. Bogatina, I.K. Yanson, B.I. Verkin, A.G. Batrak: Sov. Phys. JETP
   38, 1162 (1974)
8 Y. Skarlatos, R.C. Barker, G.L. Haller, A. Yelon: Surf. Sci. 43, 353
   (1974)
9 P.K. Hansma, R.V. Coleman: Science 184, 1369 (1974)
10 M.G. Simonsen, R.V. Coleman, P.K. Hansma: J. Chem. Phys. 61, 3789 (1974)
11 B.F. Lewis, M. Mosesman, W.H. Weinberg: Surf. Sci. 41, 142 (1974)
12 J. Klein, A. Leger, M. Belin, D. Dufourneau, M.J.L. Sangster: Phys. Rev.
   B 7, 3336 (1973)
13 Y. Skarlatos, R.C. Barker, G.L. Haller, A. Yelon: J. Phys. Chem. 79,
   2587 (1975)   .
14 J.D. Langan, P.K. Hansma: Surf. Sci. 52, 211 (1975)
15 C.S. Korman, R.V. Coleman: Phys. Rev. B 15, 1877 (1977)
16 K.P. Roenker, T.P. Graham, R.G. Keil: to be published
17 In addition to the other papers in this symposium, see two recent reviews,
   ref. 3, and N.M. Brown, D.G. Walmsley: Chem. Br., 92 (1976)
18 D.G. Walmsley, I.W.N. McMorris, N.M.D. Brown: Solid St. Commun. 16, 663
   (1975)
19 J.R. Kirtley, P.K. Hansma: Phys. Rev. B 12, 531 (1975)
20 G. Burrafato, G. Faraci, G. Giaquinta, N.A. Mancini: J. Phys. C 5, 2179
   (1972)
21 E.L. Wolf: Solid State Phys. 30, 1 (1975), and reference there in
22 J. Kirtley, D.J. Scalapino, P.K. Hansma: Phys. Rev. B 14, 3177 (1976)
23 S.L. Cunningham, W.H. Weinberg, J.R. Hardy: Bull. APS 22, 256 (1977); to
   be published
24 S.J. Niegowski, Ind. Eng. Chem. 45, 632 (1953); Sewage Ind. Wastes 28,
   1266 (1956)
25 P.S. Bailey, S.S. Bath, J.B. Ashton: Adv. Chem. Ser. 21, 143 (1959)
26 H.R. Eisenhauer: Water Pollut. Control Fed. 40, 1887 (1962)

Application of IETS to the Study of Adhesion[1]

H.W. White, L.M. Godwin, and T. Wolfram

Department of Physics, University of Missouri
Columbia, MO 65201, USA

ABSTRACT

The interface for most metal-adhesive bonds is that of an adhesive molecule
bonded to an oxide of the metal, rather than to the metal. The detailed
nature of the microscopic bondline between the oxide layer and the first
adhesive monolayer is difficult to investigate using conventional ultrasonic
and IR techniques.

A program to study the interface between a commercially important, high
performance adhesive and an aluminum oxide layer using IETS techniques has
been initiated. The epoxy system chosen is a mixture of the two molecular
compounds diamino diphenyl sulfone (DPS) and tetraglycidycl 4,4' diamino
diphenyl methane (DPM).

In our preliminary investigation IETS spectra have been obtained for each
molecule and for a mixture of the two. The DPS spectra show some evidence
for an aging effect, possibly associated with chemisorption or polymerization.
Identification of the peaks is aided by comparing peak location with IR
measurements and with the vibrational energies calculated from molecular
force constant data.

Experiments are being carried out to investigate the effects of water on
the oxide-adhesive interface using deuterium oxide. Further experiments
planned include those to investigate heat treatment and hydrothermal aging
of adhesive bonds.

1. Introduction

This paper discusses some of the work done on adhesive molecules at the
University of Missouri utilizing inelastic electron tunneling as a spectroscopic tool.

Figure 1 shows a schematic of a typical adhesive bond between two pieces
of aluminum metal. The adhesive is in contact with the aluminum oxide rather
than the metal since an oxide quickly forms on aluminum when it is exposed to
air or cleaned with etching solutions prior to bonding. In commerical applications the oxide thickness is typically 100 to 200 angstroms. There is
a fantastic need for information regarding the microscopic interface between
the adhesive and the oxide. It is difficult, however, to obtain the information using conventional methods. Ultrasonic techniques give critically
needed information about the bulk properties of the adhesive but give little
information about the interface region. The wavelengths in ultrasonic studies
are considerably longer than the several angstroms associated with the thickness of the interface layer.

The chemical nature of the interface is very important to the integrity and service life of a bond. The dashed line in Fig. 1 shows a typical fracture line. Most of the fracture line (actually a surface) is not at the interface; however, it often happens that the fracture line <u>originates</u> at the interface region. BLACK and BLOMQUIST [1] have studied several metal adhesive systems and find that when they are exposed to a heat treatment the shear strength decreases significantly, often by more than a factor of two. They concluded that the primary reason for this decrease in shear strength is due to catalytic action at the bondline.

Hydrothermal aging is believed to be one of the most important causes for bond failure in service. Studies have been done on the bulk properties in order to determine how water which permeates a bond contributes to its degradation. The role of the water at the interface is considered to be very important, but to date there have been no good means of studying its effect.

The fact that an adhesive interface consists of a molecular layer bonded to an oxide suggests the possibility of inelastic electron tunneling spectroscopy as a useful tool. Our programs in both IETS and adhesives are relatively new and, as a consequence, most of the information presented is preliminary.

The objective of this program is to assess the feasibility of utilizing IETS to monitor the chemical state of an adhesive/adherend interface of the type encountered in the adhesive bonding of aluminum components. The goal

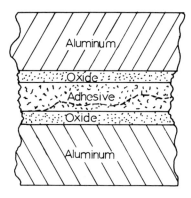

Fig.1   Schematic of an adhesive bond between two pieces of aluminum

of the project is to determine what changes occur at an adhesive bondline during thermal curing and in bond degradation resulting from hydrothermal aging. Through such studies we hope to determine some of the important mechanisms of adhesive bond failure. The high performance Hercules 3501 epoxy system, distributed by Hercules, Inc. [2] was chosen for this study.

2. Experimental

The two molecules which form the epoxy system are shown in Fig. 2. The upper one is tetraglycidycl 4,4' diamino diphenyl methane (DPM). It has four identical arms, each with an epoxy ring. The other molecule shown is diamino diphenyl sulfone (DPS) which is the curative for the DPM epoxy molecule. The molecules are not planar. The atoms in the DPS molecule lie in two planes

Tetraglycidyl 4,4′ diaminodiphenyl methane
DPM

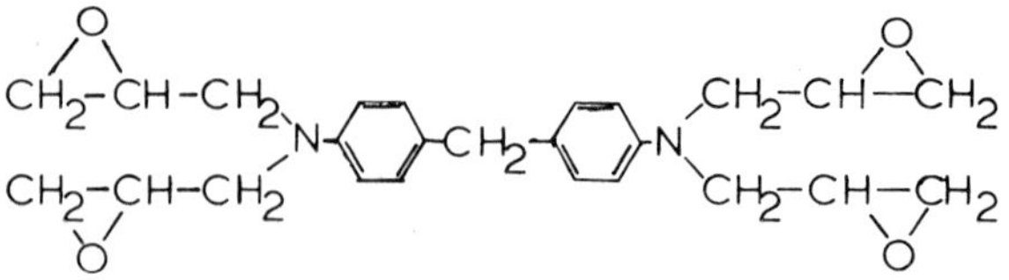

Diaminodiphenylsulfone
DPS

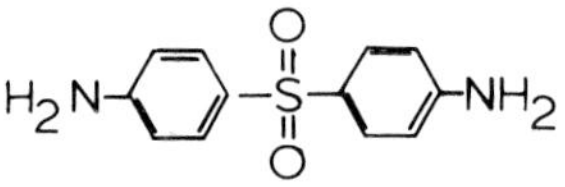

Fig.2 Schematics of the two molecules in the Hercules 3501 epoxy system

whose intersection is a line in the plane defined by the two oxygens and the sulphur. The DPM molecules are even more 3-dimensional. They form dimers relatively easily even at room temperature. Spectra must therefore be interpreted with an awareness of the fact that one or more of the arms may be joined to other molecules.

The cross-link reaction for the DPM and DPS molecules is shown in Fig. 3. In the cross-link molecule there is an OH group on each arm. The arms in the cross-linked structure link with other arms in a somewhat disorganized fashion. The result is a very 3-dimensional structure. The methane group and rings are incorporated into the matrix.

Bonding to the aluminum oxide probably occurs through the OH group, in much the same way in which phenol bonds to an aluminum oxide film, namely, as $C_6H_5O^-$ [3]. The OH groups make this molecule very polar and thereby susceptible to water permeation. The presence of water on the substrate could adversely affect the bonding properties.

Aluminum/aluminum oxide/dopant/lead tunnel junctions were fabricated, where dopant refers to the molecules on which IETS spectra are to be obtained. The aluminum electrode was first evaporated at 10^{-6} Torr. A glow discharge was then used to form the oxide layer. A liquid doping technique was used to deposit approximately one monolayer of the molecules on the oxide. The process involves making a dilute solution of the molecular species to be studied, placing a drop of the solution on the electrode, and spinning off the excess. A lead counter electrode was then evaporated over the molecules.

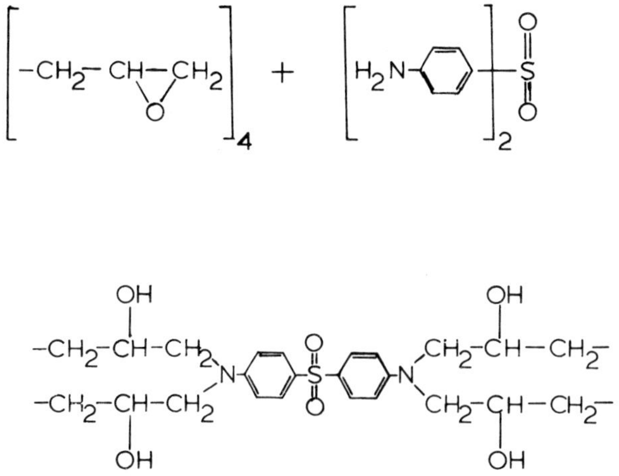

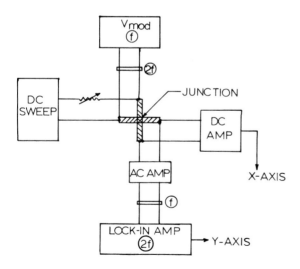

Fig.3 The cross-link reaction for the Hercules 3501 epoxy system

Fig.4 A block diagram of the inelastic electron tunneling spectrometer

The DPM resin was not easy to dissolve. Methyl ethyl ketone (mekol) and tetrahydrofuran (THF) worked best. These two solvents, and to a lesser extent chloroform, dissolved the DPS curative.

Figure 4 is a block diagram of the spectrometer system. The junction is cooled to 4.2 K. A DC bias across the junction is slowly swept from 50 to 500 millivolts. An AC signal of 1000 Hz and approximately 1 millivolt (rms) is applied continuously. The amplitude of the second harmonic signal as generated by the tunnel junction is recorded on an XY recorder as a function of the DC bias. The second harmonic signal is proportional to d^2V/dI^2. Peaks in this signal locate the energies of the vibrational modes as read from the bias voltage (energy) scale.

3. Results

Spectra were first obtained on each of the molecules, then for the two molecules placed together on the same junction. This work is to be followed by studying the effects of heat treatment and hydrothermal aging on the observed spectra. Figure 5 shows two curves, A and B, taken on a junction doped with

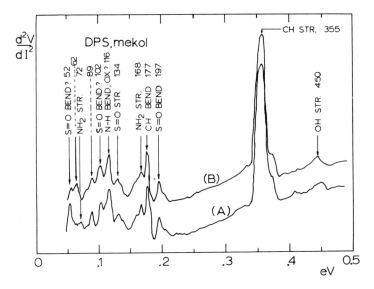

Fig.5 Two spectra taken on a junction doped with the curative DPS

DPS using mekol as a solvent. The energy range is from about 50 to 500 millielectron volts (meV), which corresponds approximately to the wavenumber range 400 to 4000 cm^{-1}. The OH stretch peak near 450 meV is observed in most IETS spectra. The CH stretch near 355 is found in all hydrocarbon spectra. The lower energy peaks for the DPS molecule were assigned by comparison with IR, other IETS data, and with normal mode frequencies calculated using literature values for the force constants. Both atomic motions and frequencies were obtained from the normal mode calculations. If an observed peak was more than 3 meV from a calculated vibrational energy a question mark was placed after its assignment label in Fig. 5. The measured locations, in meV, of the peaks are shown following the group assignment. The S=0 bend at 197 is close to the calculated value of 196 meV; however, the assignment is not certain since DPM, in Fig. 6, also shows a peak near 197. The CH band at 177, the NH$_2$ stretch at 168 and the S=0 stretch at 134 meV agree well with

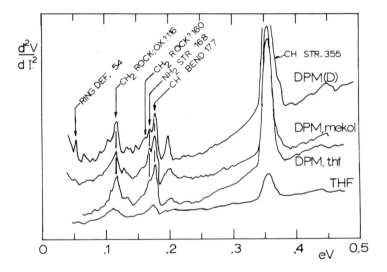

Fig.6   Spectra taken on junctions doped with the epoxy DPM

calculated vibrational energies and other data. The peak at 116 meV could be either a NH bend or an oxide phonon. Many doped junctions and nearly all undoped junctions show the presence of an oxide phonon at 117; however, the calculated vibrational energy for the NH bend is 115 meV. The peak at 102 meV was tentatively identified as an S=0 bend since this group had a calculated frequency near 98 meV. The assignment at 52 meV was made in the same manner except that the S=0 motion had a calculated frequency of 48 meV.

Until now the features discussed have been the same for both spectra A and B. In the energy range 55 to 75 meV the spectra are different. Spectrum A was taken on the junction soon after preparation. Spectrum B was taken nine days later, during which time the junction had been stored in a clean, dry atmosphere. The junction resistance remained constant at 1500 ohms. Spectrum A shows a small peak at 72 meV which is within 2 meV of the value of the normal mode energy calculated for an NH_2 stretch in DPS. Spectrum B shows no evidence for a peak at 72 meV but shows a new distinct peak at 62 meV. An aging effect has occurred. There is very little evidence in the literature of any polymerization occurring inside of a tunnel junction, but there is evidence for chemisorption. We speculate that the NH_2 groups on the ends of the DPS molecule have chemisorbed to the oxide layer.

Figure 6 shows the spectra of four separate junctions. The lower curve is the spectrum for a junction doped with only the solvent THF. It serves as a background spectrum. Both THF and mekol spectra show a small peak near 174 meV which is undoubtedly due to CH bend modes and a small peak near 117 meV which is probably due to an oxide phonon. The second curve from the bottom, labeled "DPM, THF" is for the DPM epoxy molecule dissolved in the THF. Again, there is a large CH stretch peak located at 355 meV. The third curve labeled "DPM, mekol" shows essentially the same features as the second. The peak at 177 meV is probably a CH bend. The peak at 168 meV is probably due to NH_2 stretch modes. CH_2 rock modes are expected to be observed near 116 meV; however, the peak at that location could be due, at least in part, to an oxide phonon expected near 117 meV.

The curve labeled "DPM(D)" is a spectrum for DPM which had been deuterated by an exchange reaction before placing in the junction. One of the goals of this project is to study the role of water at the interface of an adhesive bond. Water can be diffused into a junction and its presence can be monitored by using D_2O as tag molecules [4]. The "DPM(D)" spectra was taken to provide information on the size and location of peaks associated with a deuterated DPM molecule. The spectrum has peaks not observed in the other spectra. We must be cautious in our interpretation of this spectrum since some structural changes may have occurred during the deuteration process.

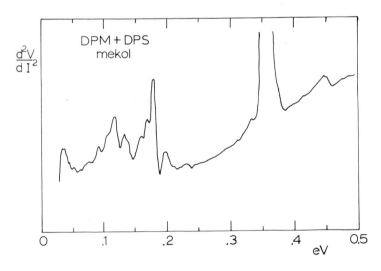

Fig.7 A spectrum on a junction containing both epoxy (DPM) and curative (DPS) molecules

Figure 7 shows a spectrum for a junction containing both DPS and DPM molecules. Its features were similar to those in spectrum A of the DPS molecule. One reason for this similarity may be due to the fact that the signature for the DPS molecule was always much stronger than that for the DPM molecule. Obtaining good noise-free junctions was considerably more difficult for the DPM molecule than for the smaller DPS. We are currently trying to improve spectral resolution and do some heat treatment and aging studies on the combined system.

4. Summary

IETS spectra have been obtained for the components of the high performance commercial adhesive, Hercules 3501. This epoxy system consists of two molecular components; diamino diphenyl sulfone (DPS) and tetraglycidyl 4,4' diamino diphenyl methane (DPM). Vibrational modes have been assigned by comparison with computer calculations and existing infrared optical spectra. Some evidence for an aging effect has been observed for the adsorbed DPS. This effect appears as a dramatic change in low frequency vibrational modes and may be associated with the formation of hydrogen bonding of the NH_2 groups to the oxide layer.

Acknowledgements

The authors express their appreciation to D.O. Thompson for suggestion of the problem, to D. Kaelble, R.C. Jaklevic and J. Lambe for helpful discussions, and to H. Kim for providing a program for calculation of vibrational mode frequencies. We thank P.K. Hansma for help in equipment design.

[1]Supported in part by ARPA/AFML Contract No. F336615-74-C-5180

References

1 J.M. Black, R.F. Blomquist: Ind. Eng. Chem. 50, 918 (1958)
2 Hercules, Inc., P.O. Box 98, Magna, Utah 84044
3 B.F. Lewis, W.M. Bowser, J.L. Horn, Jr., T. Luu, W.H. Weinberg: J. Vac. Sci. Technol. 11, 262 (1974)
4 See the paper by R.C. Jaklevic and M.R. Gaerttner in this proceedings

III. Theoretical Aspects of Electron Tunneling

Theoretical Interpretation of IETS Data

J. Kirtley

Department of Physics and Laboratory of Matter, University of Pennsylvania
Philadelphia, PA 19174, USA

ABSTRACT

We first discuss the initial theoretical work on IETS by SCALAPINO and MARCUS, which included a dipole potential of the molecule and its image in the tunneling barrier potential and calculated the excess tunneling current due to this potential within the WKB approximation for infrared active modes. The theory obtained the proper intensity for the O-H stretch mode on alumina and predicted that the orientation of the molecules on the surface could be inferred from the tunneling spectrum, since only dipole moments oscillating perpendicular to the surface coupled to the tunneling electron. JAKLEVIC and LAMBE extended this theory to Raman modes by including the bond polarizabilities, and predicted that Raman modes should be observable in IETS with intensities about 2-5 times smaller than the infrared modes. KLEIN et al. used a transfer Hamiltonian formalism to show that LO as well as TO phonons could be observed in MgO using IETS. KIRTLEY, SCALAPINO, and HANSMA applied the transfer Hamiltonian formalism to molecular vibrations in IETS. They replaced the dipole approximation with a set of partial charges located on the atoms in the molecule, and allowed for off-axis electron scattering. In this theory the local nature of the electron-molecular interaction weakened the orientation and symmetry selection rules. It predicted that the Raman modes should be observable even neglecting bond polarizabilities, and that optically forbidden modes, although weak, may be observable. We discuss the formal difficulties involved with the theory at present, compare our predictions with experiment, and try to point the way toward further development of the theory of intensities in IETS.

Inelastic electron tunneling spectra contain a great deal of information about the structure and chemical configuration of molecular species in the tunneling region. On one hand, the vibrational mode energies as measured using IETS are well understood, at least in the sense that the voltage of a peak in d^2V/dI^2 in the tunneling characteristic gives the vibrational frequency of a normal mode of the molecule through the relation $eV = \hbar\omega$ [1]. At finite temperatures there is thermal smearing of the peak [1,2], and the modulation technique commonly used contributes additional smearing [3] and (in the presence of a superconducting electrode) some shifts of the vibrational mode peak voltages [4]. But the connection between the experimental peak voltages and the vibrational mode energies is straightforward. The unraveling of information from the normal mode frequencies as measured, although quite promising, is less straightforward, but much work in this area has been done in the support of the optical spectroscopies, and I don't wish to treat that very complex subject here.

On the other hand, the intensities of the peaks in IETS are much less well understood. Spectroscopic intensities are notoriously difficult to calculate in general, but there is much to be learned from a detailed theory: (1) an understanding of the rules governing intensities will tell us which modes are

likely to appear strongly for a given molecule, easing identification of surface species; (2) the degree of breaking of these rules may indicate the extent of the perturbation of the molecule by the surface; (3) it may be possible to infer the orientation of molecules on the surface by comparing selected mode intensities; (4) finally, it may be possible to infer the chemical configuration of molecules on the surface by a careful study of peak intensities.

It will be useful to give first a very brief review of elastic tunneling
theory [5]. Historically the first approach to calculating tunneling probabilities (Fig. 1a) involves calculating the solution to the time independent
Schroedinger equation for a given barrier potential which has the form of
incoming and reflected plane wave (or Bloch) states on one side of the barrier,
as well as transmitted states on the other side. By matching wave functions
and their normal deviations at the boundaries, one can calculate transmission
probabilities $D(\vec{k}_i, \vec{k}_f)$ by taking the ratio of the transmitted to the incident
fluxes. Tunneling currents are then obtained by summing over incoming and
outgoing momenta. If it is assumed that momentum parallel to the interface
is conserved, it can be shown that

$$j_e = \frac{2e}{\hbar} \int dE \left(f(E) - f(E + eV) \right) \int \frac{d^2 k_{\parallel}}{(2\pi)^2} \left. D(\vec{k}_i, \vec{k}_f) \right|_{k_{i\parallel} = k_{f\parallel}} \tag{1}$$

This approach is exact within the limits of the approximations to the potentials that are made. Although somewhat cumbersome when applied to the
case of complex potentials in three dimensions, it has been used for many
calculations of elastic tunneling processes. Applications of this "exact"
approach to inelastic tunneling have been made by BRAILSFORD and DAVIS [6],
CAROLI et al. [7], and T. E. FEUCHTWANG [8].

A second approach, which was first applied to tunneling by BARDEEN [9] to
help explain the observation by GIAEVER of gap structure in the current-
voltage characteristics of normal-superconducting tunnel junctions, is illustrated in Fig. 1b. This approach calculates tunneling rates by using first-
order time dependent perturbation theory. Although originally developed as
a many body theory, it is somewhat easier to present in a single particle
picture. The initial electron state is localized on the left side of the
tunneling barrier by extending the barrier potential such that $z_R \to \infty$. The
final state is similarly localized by setting $z_L \to -\infty$. If the WKB approximation for the wave functions is used, the initial wave functions inside and
outside the barrier are given by:

$$\psi_i^{out} \propto k_z^{-\frac{1}{2}} e^{i(k_x x + k_y y)} \sin(k_z z + \gamma)$$

$$\psi_i^{in} \propto |k_z|^{-\frac{1}{2}} e^{i(k_x x + k_y y)} e^{-\int_{z_L}^{z_R} |k_z| \, dz} \tag{2}$$

where

$$|k_z| = \sqrt{\frac{2m}{\hbar^2} \left[V(z) - \varepsilon_{kz} \right]}$$

and $V(z)$ is the barrier potential. Similar results are obtained for the
final state wave functions.

81

To treat elastic tunneling, we take the solution of the time-dependent Schroedinger equation to be a linear combination of initial and final state wave functions:

$$\psi(t) = a(t)\ \psi_i\ e^{-i\omega_0 t} + \sum_f b_f(t)\ \psi_f\ e^{-i\omega_f t} \tag{3}$$

Plugging this into the Schroedinger equation and solving to first order for $b_f(t)$ leads to the standard Fermi's Golden Rule result for the transition rate/unit time:

$$\omega_{if} = \frac{2\pi}{\hbar}\ |M_{if}|^2\ \delta\ (\omega_0 - \omega_f) \tag{4}$$

with

$$M_{if} = -\frac{\hbar^2}{2m} \int dx \int dy \left[\psi_i^\star \frac{\partial}{\partial z} \psi_f - \psi_f \frac{\partial}{\partial z} \psi_i^\star \right]\Bigg|_{z = \text{const}} \tag{5}$$

where the integral can be evaluated anywhere inside the barrier. For WKB type wave functions the matrix element is given by:

$$M_{if} \propto e^{-K(z_L - z_R)} \tag{6}$$

with

$$K = \sqrt{\frac{2m}{\hbar^2}\ (V_0 - \epsilon_{kz})} \qquad \text{where } V_{\cdot 0} = V_{00} - |eV|/2 \tag{7}$$

and V_{00} = average barrier height at zero bias.

To obtain the total (elastic) current we sum over initial and final allowed states. Many body effects can be inserted as tunneling densities of states to obtain:

$$j_e = \frac{4\pi e}{\hbar} \sum_{k_i} \sum_{k_f} |M_{if}|^2 \left[f(\epsilon_i) - f(\epsilon_f + eV) \right] N_i(\epsilon_i) N_f(\epsilon_f + eV)\ \delta(\epsilon_i - \epsilon_f) \tag{8}$$

Note that in (8) the density of states functions $N_i(\epsilon_i)$ appear, while they do not appear in (1). This basic approach, which is called the transfer Hamiltonian formalism, has been applied in several theories of inelastic tunneling; 2 district applications will be discussed here. The drawbacks of the transfer Hamiltonian approach have been pointed out by several authors [7,8,9], and will be discussed by T. E. FEUCHTWANG later in this conference. Let me just say that the transfer Hamiltonian approach is an approximation in that: 1) it uses perturbation theory in a non-equilibrium situation, 2) it defines 2 separate sets of states for the left and right electrodes which are not orthogonal and hence are not a complete set of states for the full system. However, this approach is sufficiently simple to allow the use of realistic molecular potentials, and has been applied with great success (e.g. to describe gap structure, Josephson effects, and phonon structure) in M-I-M tunneling junctions. It can be shown that the transfer Hamiltonian approach gives the same results as more sophisticated models in the limit of thick barriers.

The first theoretical treatment of inelastic electron tunneling[1] was

---

[1]This talk is intended as a brief review of a fairly complex body of theoretical work. To review this work in some depth in a limited space I have been

forced to slight the discussions of (in chronological order) C.B. DUKE [12, 13], BRAILSFORD and DAVIS [6], CAROLI et al. [7], KLEIN and LEGER [3], and T. E. FEUCHTWANG [8]. The interested reader is herby referred to their work.

given by SCALAPINO and MARCUS [11]. They assumed that the tunneling electron-molecular potential due to the molecular dipole and its image in the metal surface was of the form:

$$U_I(z) = \frac{2ep_z z}{(z^2 + r_\perp^2)^{3/2}} \tag{9}$$

(see Fig. 2). They added (9) to the barrier potential and calculated the excess tunneling current due to this potential within the transfer Hamiltonian formalism. Following the calculation presented above for the elastic current, and using WKB type wave function (2) they found that the matrix element M_{if} was given by (4) as before, but with

$$K = \sqrt{\frac{2m}{\hbar^2} (V_0 + U_{int} - \varepsilon_{kz})} \tag{10}$$

Setting $V_0 - \varepsilon_{kz} = \phi_0$, a constant, and expanding to lowest order in U_{int}/ϕ_0, they found

$$M_{if} \propto \left[1 + \left(\frac{2m}{\phi}\right)^{1/2} \frac{ep_z}{\hbar\ell} g\left(\frac{r_\perp}{\ell}\right)\right] e^{-\left(\frac{2m\phi}{\hbar^2}\right)^{1/2}\ell}$$

where $\ell = |z_L - z_R|$, $g(x) = \frac{1}{x} - \frac{1}{(1+x^2)^{\frac{1}{2}}}$

   In this expression p_z is taken to be the expectation value of the dipole moment operator between the ground state and the first excited state of the vibrational mode in question (at these low temperatures, all of the molecules are initially in their ground state). For inelastic tunneling electrons only flow in one direction, and the sum over initial and final states (8) becomes:

$$j_i = \frac{4\pi e}{\hbar} \sum_{k_i} \sum_{k_f} |M_{if}|^2 \left[f(\varepsilon_i)\right] \left[1-f(\varepsilon_f+eV)\right] N_i(\varepsilon_i) N_f(\varepsilon_f+eV) \delta(\varepsilon_i-\varepsilon_f-\hbar\omega_m) \tag{11}$$

If we assume that the momentum of the electron parallel to the interface is conserved, it is easy to show that the ratio of the inelastic to the elastic conductances is given simply by the square of the ratio of the change in the elastic matrix element divided by the total elastic matrix element:

$$\frac{\frac{dj_i(r)}{dV}}{\frac{dj_e}{dV}} = \frac{2m}{\psi} \left(\frac{e}{\hbar\ell}\right)^2 |<1|p_z|0>|^2 g^2(r_\perp/e) \theta(V - \frac{\hbar\omega_m}{e}) \tag{12}$$

$$\theta(x) = \begin{cases} 1, x > 0 \\ 0, x < 0 \end{cases}$$

In this expression the θ function allows the inelastic channel to open up only when the electrons have enough energy to excite the vibration $eV > \hbar\omega_m$.

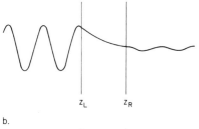

z_L z_R

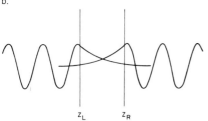

z_L z_R

METAL   INSULATOR   METAL

Fig.1 Schematic drawings of
the wavefunctions across a
one-dimensional tunneling
barrier. In the "exact" calculation (1a) an incident state
from the left decays exponentially into the barrier and has
some probability of appearing
on the right. In the transfer
Hamiltonian formalism (1b),
initial and final wave functions on either side of the
barrier "transfer" from one
side to the other at a finite
rate

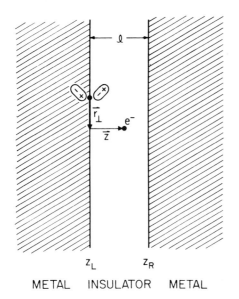

z_L z_R

METAL   INSULATOR   METAL

Fig.2 Geometry used in the
theory of SCALAPINO and MARCUS.
The tunneling electron at
$\vec{r} = \vec{r}_\perp + \vec{z}$ interacts with the
vibrating molecule through a
dipole potential due to the
molecule and its image in the
metal electrode

To obtain the total inelastic conductance due to one vibrational mode of the impurity we integrate (12) over $r_\perp$ using

$$\int_{r_0}^{\ell} dr_\perp \ r_\perp \ g^2 \left(\frac{r_\perp}{\ell} \right) \simeq 2\pi_\ell^2 \ \ell n \left(\left| \frac{\ell}{r_0} \right| \right) \tag{13}$$

To make this integration converge we must use two cutoffs: r_0 and ℓ. Since the ratio of these cutoffs appears only in a log term, their exact values are not extremely critical, but their presence is a weakness of this approach. We then multiply by N impurities per unit area and sum over all vibrations of each impurity to obtain:

$$\frac{dj_i/dV}{dj_e/dV} = \frac{4\pi Nme^2}{\phi \hbar^2} \ \ell n(\ell/r_0) \ \sum_m |{<}m|p_z|0{>}|^2 \ \theta \left(V - \frac{\hbar\omega_m}{e} \right) \tag{14}$$

For reasonable values of ℓ, r_0, ϕ, N, m, and p_z the calculated magnitude of the conductance jump due to a monolayer of OH ions is ~1%, which agrees with experiment. The theory also predicts that only vibrational modes with a dipole moment perpendicular to the surface cause inelastic transitions. Further, it predicts that only vibrational modes involving a change in the dipole moment of the molecule are observed using IETS, with an intensity proportional to p_z^2. This is the same proportionality factor that appears in infrared absorption, so one might expect a strong correlation between infrared and tunneling intensities. Unfortunately, while in general the strong IR modes are strong in IETS, the correlation is not exact. Further, vibrational modes appear in IETS that have no net dipole moment. JAKLEVIC and LAMBE [1] extended the theory of SCALAPINO and MARCUS to include the affects of the interaction of the tunneling electron with the induced dipole moment of the molecule through the bond polarizability. This term couples to Raman active modes, which have a change in the polarizability associated with a given vibration, but not necessarily a change in the net dipole moment. The interaction energy, again including effects of the nearest image of the dipole, is

$$U_I^R = \frac{-4\pi e\alpha z^2}{(z^2+r_\perp^2)^3} \tag{15}$$

Inserting this additional term into the barrier potential and proceeding as before, we obtain:

$$|M_{if}^R| \ \alpha \ \left[1 + \left(\frac{2m}{\phi} \right)^{\frac{1}{2}} \frac{\alpha e^2}{4\hbar \ell^3} t \ (\ r_\perp/\ell \ ) \right] e^{- \left(\frac{2m\phi}{\hbar^2} \right)^{\frac{1}{2}} \ell}$$

where $\quad t(x) = \frac{1}{x^2} \left[\frac{1-x^2}{1+x^2} + \frac{1}{x} \tan^{-1} \left(\frac{1}{x} \right) \right]$

Then

$$\frac{\dfrac{dj_i^R}{dV}}{\dfrac{dj_e}{dV}} = \frac{4\pi mNe^2}{\phi \hbar^2} \ \frac{e^2}{16\ell^2} \ \int_{r_0}^{\ell} t^2 \left(\frac{r_\perp}{\ell} \right) r_\perp^2 \ dr_\perp \ \sum_m |{<}m|\alpha|0{>}|^2 \theta \left(V - \frac{\hbar\omega_m}{e} \right) \tag{16}$$

In this case $t(x)$ is strongly divergent as $x \to 0$ and the integral is very dependent on the value of the cutoff r_0. The reason for this divergence is that the polarizability α is taken to be independent of $r_\perp$, while physically α must get small as $r_\perp \to 0$. Nevertheless, estimates of the size of the conductance step for reasonable values of $r_\perp$, ℓ, N, ϕ, and α give magnitudes of (.1 - .5%) as opposed to 1% for the electron-dipole interaction. Therefore both Raman and infrared active modes should be observable in IETS.

The theory of SCALAPINO and MARCUS [11] as extended by JAKLEVIC and LAMBE [1] works well for calculating the orders of magnitude of the tunneling intensities. However, there are some difficulties: 1) It assumes that the momentum parallel to the interface is conserved in the tunneling process. Strictly speaking this is not the case, and several interesting effects occur when off-diagonal scattering is allowed. 2) In calculating the tunneling current as a function of $r_\perp$, it is implicity assumed that the electronic wavefunction is localized on a scale small with respect to the interaction distance. This assumption leads to divergent integrals unless cutoffs (in $r_\perp$) are introduced. 3) The molecular potential is taken for the molecule as a whole, although the electrons scatter over short distances ($\sim$2Å) and could therefore be expected to "see" the molecular structure.

In order to avoid these difficulties, KIRTLEY, SCALAPINO, and HANSMA [14] used a transfer Hamiltonian formalism with a local, although still empirical, interaction potential.

$$V_I(\vec{r}) = - \sum_i \frac{e^2 Z_i}{|\vec{R}_i - \vec{r}|} \tag{17}$$

Where $\vec{R}_i$ is the position of the ith atom in the molecule and $Z_i e$ is the partial charge associated with the ith atom. The partial charges result from unequal sharing of the charge density of the bonding electrons in the molecule, and can either be taken as empirical parameters or assigned the values of the dipole derivatives of the bonds. As the molecule vibrates the atomic positions move to first order as

$$\vec{R}_i = \vec{R}_{io} + \delta\vec{R}_i e^{i\omega t} \tag{18}$$

where the set of $\delta\vec{R}_i$'s depend on the vibrational mode of interest. In the Golden rule formalism, the only part of the total interaction term $V_I(\vec{r})$ that connects initial and final states with different energies is the component that oscillates at frequency ω. Expanding (17) to first order in $\delta\vec{R}_i$, we find

$$V_I(\vec{r},\omega) = - \sum_i e^2 Z_i \, \delta\vec{R}_i \cdot \vec{\nabla}_i \left(\frac{1}{|\vec{R}_i - \vec{r}|} \right) \tag{19}$$

Following the transfer Hamiltonian recipe, we define initial and final wave functions in the WKB approximation (2). But instead of using $V_I(r,\omega)$ in the definition of K and using the full Hamiltonian to transfer electrons across the barrier, (as was done by SCALAPINO and MARCUS) we use only the zeroth order Hamiltonian

$$H_0(\vec{r}) = \frac{p^2}{2m} + \left. \begin{array}{c} V \\ 0 \end{array} \right\} \begin{array}{c} z_R < z < z_L \\ \text{otherwise} \end{array} \tag{20}$$

to calculate K, and use $V_I(r,\omega)$ to transfer the electron across the barrier. This turns out to be a much simpler approach for a complicated potential [15]. The matrix elements are then given by:

$$M_{if} = <\psi_f|V_I(r,\omega)|\psi_i> \tag{21}$$

The calculations of the matrix elements are somewhat complex but quite straightforward. We take into account the infinite series of images of the moving charges in the two metal surfaces as well as allowing for the dielectric screening of the potential by the oxide itself. If we separate out the contribution from each atom in the molecule, the contribution to the total matrix element of the nth atom at position $\vec{r} = a\hat{z} + b\hat{x} + c\hat{y}$ with partial charge Z_n and displacement $\delta\vec{R}_n = \sum_m \delta_{nm}e^{i\omega t}\hat{t}_m$ is given by

$$M_{if}^{nm} = \frac{\pi e^2 Z_n \delta_{nm}\xi_z e^{-|K_{zi}|\ell}e^{i\alpha_x b}e^{i\alpha_y c}}{\epsilon L^3} \frac{\alpha_j}{\alpha_\parallel}\left\{ \frac{e^{\alpha_\parallel a} + e^{-\alpha_\parallel a}}{e^{2\ell\alpha_\parallel} - 1} \left[\frac{e^{(\alpha_z-\alpha_\parallel)\ell}-1}{\alpha_\parallel - \alpha_z}\right.\right.$$

$$+ \left.\frac{e^{(\alpha_z+\alpha_\parallel)\ell}-1}{\alpha_\parallel + \alpha_z}\right] + \frac{1}{\alpha_\parallel - \alpha_a}\left[e^{(\alpha_z-\alpha_\parallel)\ell}\left(e^{\alpha_\parallel a} \pm e^{-\alpha_\parallel a}\right) - \left(e^{\alpha_z a} \pm e^{-\alpha_\parallel a}\right)\right] \tag{22}$$

$$\pm \frac{1}{\alpha_\parallel + \alpha_z}\left[e^{\alpha_z a} - e^{-\alpha_\parallel a}\right]\Bigg\}$$

where
$$\alpha_x = k_{xi} - k_{xf}$$

$$\alpha_y = k_{yi} - k_{yf}$$

$$\alpha_z = |K_{zi}| - |K_{zf}|$$

$$\alpha_\parallel = (\alpha_x^2 + \alpha_y^2)^{1/2}$$

$$|K_z| = \left(\frac{2m}{\hbar^2}\right)^{\frac{1}{2}}\left(V - \frac{\hbar^2 k_z^2}{2m}\right)^{\frac{1}{2}}$$

$$\xi_z = \left(\frac{k_{zi}k_{zf}}{|K_{zi}|\,|K_{zf}|}\right)^{\frac{1}{2}}$$

In this equation the upper sign is taken and $\alpha_j = \alpha_\parallel$ if $\hat{j} = \hat{z}$ (vibrations perpendicular to the metal surfaces). The lower signs are taken and $\alpha_j = \alpha_x$, α_y if $\hat{j} = \hat{x}, \hat{y}$ (motion parallel to the metal surfaces). We then sum over i, j to find the total matrix element, which is in general a function of incoming and outgoing energies and directions

$$|M_{if}|^2 = | \sum_{nm} M_{if}^{nm}|^2 = G (\varepsilon_i, \varepsilon_f, \hat{k}_i, \hat{k}_f) \tag{23}$$

Since IETS is done at low temperatures, we can take the low temperature limits of the Fermi functions in (11). Changing the wavevector sum to an integral, we find:

$$j_i = \frac{8\pi e}{\hbar} \left(\frac{L}{\pi}\right)^6 \left(\frac{m}{\hbar^2}\right)^3 \int d\Omega_i \int d\Omega_f \int_0^\infty d\varepsilon_i \, \varepsilon_i^{\frac{1}{2}} \int_0^\infty d\varepsilon_f \, \varepsilon_f^{\frac{1}{2}} \tag{24}$$

$$G(\varepsilon_i, \varepsilon_f, \hat{k}_i, \hat{k}_f)(1 - \theta(\varepsilon_i - \varepsilon_{Fi}))\theta(\varepsilon_f - \varepsilon_{Ff} + eV) \, \delta(\varepsilon_i - \varepsilon_f - \hbar\omega)$$

where ε_i, ε_{Ff} are the Fermi energies of the metal electrodes.

Evaluating (24) would be a formidable task, considering the complexity of the matrix elements (22). Fortunately, the quantity of experimental interest is the second derivative of (24) with respect to voltage. The step functions become delta functions upon differentiation and the effect of the delta functions is to put the initial and final electrons on the Fermi surface. This is just as we expect, since the onset of the tunneling path occurs when the most energetic electrons find the first available open states as the voltage is increased. We obtain finally

$$\frac{d^2 j_i}{d(eV)^2} = \frac{8\pi e}{\hbar} \left(\frac{L}{\pi}\right)^6 \left(\frac{m}{\hbar^2}\right)^3 \sqrt{\varepsilon_{Fi}} \ \sqrt{\varepsilon_{Ff} - eV} \int d\Omega_i \int d\Omega_f \, G(\varepsilon_F, \varepsilon_F + eV, \hat{k}_i, \hat{k}_f) \delta(\hbar\omega - eV) \tag{25}$$

In many cases the integration over ϕ_i can be done analytically. The other three integrations are done numerically.

Absolute values for the change in conductance due to a given vibrational mode can be obtained from (25) by integrating over voltage (removing the delta function), and multiplying by n, the density of impurities/area. When reasonable values for Z_i, δ, n, ε, ℓ, a, and V are used, the theory predicts a change of conductance of .5% for a monolayer of OH⁻ ions, in reasonable agreement with experiment.

One observation that has long puzzled investigators in IETS is that the peak intensities are different for opposite bias polarities: a given vibrational peak will appear larger for Al biased negative with respect to Pb than for Al biased positive (Fig. 3). The theory shows that, crudely speaking, the reason for this asymmetry is that in one direction the electrons tunnel across most of the barrier before losing energy, while in the other direction they tunnel after losing energy. Since more energetic electrons are more likely to tunnel, the inelastic process is more likely to occur for electrons tunneling from the Al through the oxide and scattering from impurities on the oxide surface into the Pb. This happens when the Al is biased negative. Figure 4 shows that the theory agrees fairly well with experiment for the intensity asymmetry when a small correction is made for the elastic asymmetry.

Another very interesting observation that can be made is that the individual

terms if the total matrix element (22) interfere with each other either con-
structively or destructively. This interference tends to weaken both the
orientation and symmetry selection rules. For example, the matrix element
for an OH^- ion has terms proportional to $\cosh(\alpha_{||}a)$ for the internuclear

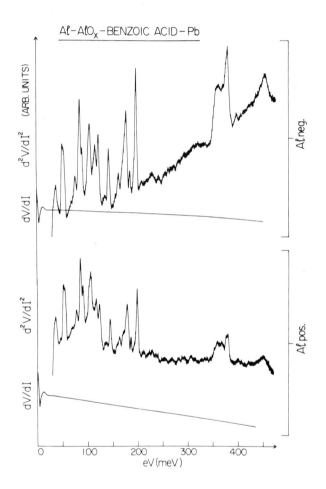

Fig.3 Experimental traces showing the asymmetry of the peak intensities
(d^2V/dI^2) and differential resistance (dV/dI) for a benzoic acid doped
$Al-AlO_x-Pb$ tunneling junction for opposite bias voltages

axis lying perpendicular to the interface but these terms are proportional to
$\sinh(\alpha_{||}a)$ for the OH^- parallel to the interface (where $\alpha_{||}$ is the parallel
momentum transfer in scattering and a is the metal electrode to atom distance).
Therefore the orientation "selection rule" (which in the theory of SCALAPINO
and MARCUS says that vibrations normal to the interface should be stronger
in intensity than vibrations parallel to the interface) is weakened in the
presence of off normal scattering. This weakening is not too pronounced,

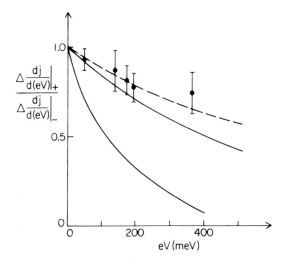

Fig.4 Plot of the ratio of the peak intensities for opposite bias voltages for benzoic acid on alumina samples, for vibrational modes of various energies. The lower solid curve is the predicted ratio if a very short range interaction is assumed [16]. The middle curve is predicted by the theory of KIRTLEY, SCALAPINO, and HANSMA; the upper (dashed) curve results from this theory if a small correction for the asymmetry of the elastic conductances is made

since calculations predict that the ratio of intensities for the two orientations should be about 10.

The interference effects also appear when we look at symmetry "selection rules". The interaction potential (18) has the form of a sum of dipoles localized on the atoms in the molecule. It could be expected, therefore, that for a molecule with a center of symmetry the IR active modes, which involve a net dipole moment, should appear more strongly in IETS than the Raman active modes, which have no net dipole moment. The complete addition or cancellation of the two classes of modes are modified by the interference effects. For example, for CO_2 oriented parallel to the interface, the matrix element is proportional to $\cos(\alpha_x b)$ for the infrared active vibration, but proportional to $\sin(\alpha_x b)$ for the Raman active mode (where α_x is the momentum transfer parallel to the internuclear axis and b is the C-O bond length). In this case the selection rule is weakened a great deal since electrons tunneling normal to the interface don't couple to vibrations parallel to the interface, and "Raman active" electrons (those scattering off normal) are favored. We calculate in this case that the Raman mode should only be about 40% weaker than the IR mode. The theory also predicts that it may be possible to observe optically forbidden transitions using IETS. These calculations show that a semi-empirical theory of intensities in IETS fits the magnitudes of the observed peaks well, accounts for the symmetry in peak heights for opposite bias voltages, and shows how the presence of off-normal tunneling can contribute to the weakening of symmetry and orientational selection rules. (As a rule of thumb, the selection rules are weakened more for 1) larger molecules, and 2) linear vs. ring molecules).

I think it is fair to say that the theoretical work I have described, although useful as a qualitative description of the inelastic tunneling process, is only a first step towards a full understanding of this new and potentially very powerful spectroscopy. Some particular weaknesses of (and possible new approaches to) the theory as it stands are: 1) the interaction between the tunneling electron and the vibrating molecule is described semi-empirically. Ab initio calculations should be made to see if these approaches are justified, amd improve on them if they are not. 2) The transfer Hamiltonian formalism is an approximation. Exact solutions, although cumbersome, may show some very interesting effects, in particular in terms of multiple scattering in the molecular-electrode interface [6]. 3) The affect of the molecule on the scattered wavefunctions themselves have not been included and could well be very important. Again, a multiple scattering theory is indicated. 4) The barrier region has been rather crudely modeled, with a trapesoidal insulating barrier surrounding the molecular potential. It could be very useful to put the scattering organic layer on the surface of the oxide, and see what happens.

Finally, let me say that there is a great deal left that these new theoretical efforts can attempt to explain. No calculation has yet been done to compare theory with the relative intensities of all the vibrational modes of a molecule measured using IETS. In particular, although the final word is not in, there is work by WEINBERG [17] and more recently COLEMAN [18] which seems to indicate that the intensities of all the ring modes of a benzene ring molecule increase and decrease in unison depending on the nature of the ligands attached and/or the orientation of the molecule on the surface. Present theory cannot explain this work.

References

1 J. Lambe, R.C. Jaklevic: Phys. Rev. 165, 821 (1968)
2 R.J. Jennings, J.R. Merrill: J. Phys. Chem. Solids 33, 1261 (1972)
3 J. Klein, A. Leger, M. Belin, D. Defourneau: 7, 2336 (1973)
4 John Kirtley, P.K. Hansma: Phys. Rev. B 13, 2910 (1976)
5 Two very complete treatments of this subject are contained in: Tunneling Phenomena in Solids, ed. E. Burstein, S. Lindquist, (Plenum Press, New York, 1969); C.B. Duke, Tunneling in Solids (Academic Press, New York, 1969)
6 A.D. Brailsford, L.C. Davis: Phys. Rev. B 2, 1708 (1970); L.C. Davis: Phys. Rev. B 2, 1714 (1970)
7 C. Caroli, R. Combescot, P. Nozieres, D. Saint-James: J. Phys. C: Solid State Phys. 4, 916 (1971); 5, 21 (1972)
8 J.E. Feuchtwang: Phys. Rev. B 10, 4121 (1974); 10, 4135 (1974); 13, 517 (1976)
9 J. Bardeen: Phys. Rev. Lett. 6, 57 (1961)
10 C.B. Duke: loc. cit., p. 207
11 D.J. Scalapino, J.M. Marcus: Phys. Rev. Lett. 18, 459 (1967)
12 C.B. Duke: loc. cit., p. 290
13 C.B. Duke, G.G. Kleiman, J.E. Stakelon: Phys. Rev. B 6, 2389 (1972)
14 John Kirtley, D.J. Scalapino, P.K. Hansma: Phys. Rev. B 14, 3177 (1976)
15 This was first pointed out by C.B. Duke in ref. 10
16 J.K. Yanson, N.I. Bogatina, B.I. Verkin, O.I. Shklarevski: Zh. Eksp. Teor. Fiz 62, 1023 (1972). [Sov. Phys.-JETP 35, 540 (1972)]
17 W.H. Weinberg: Proc. Am. Phys. Soc. 21, 281 (1976)
18 R.V. Coleman: private communication

A Model for Electron-Molecule Interaction in IETS

J.Rath and T. Wolfram

Department of Physics, University of Missouri-Columbia
Columbia, MO 65201, USA

ABSTRACT

The inelastic current in Inelastic Electron Tunneling Spectroscopy (IETS) is due to the interaction of the tunneling electron with the vibrational modes of the molecule. The intensity of IETS spectrum should depend on the details of this interaction. Previously this interaction was described by making use of fictitious 'partial charges' located on each atom of the vibrating molecule. We report here a 'first principle' approach to the calculation of relative IETS peak intensities. Initial calculations have been made for Ethylene molecule. The present calculation employs a more realistic description of the molecular charge distribution and its vibrational properties. The charge density is obtained using linear combination of atomic orbital method. The polarization vectors and frequencies for the vibrating molecule is obtained using a force constant model. The electron-molecule interaction is dependent on the gradient of the molecular charge distribution with respect to nuclear displacements and consists of nuclear, Coulomb, and exchange terms. Our results for Ethylene molecule indicate that the interaction can not be described in terms of the partial charges located on each atom.

1. Introduction

Inelastic electron tunneling spectroscopy (IETS) is a sensitive method for studying the vibrational properties of organic molecules and their interactions with oxide surfaces. In an IETS spectra there appear sharp peaks at bias voltages corresponding to the energies of the various vibrational modes of a molecule. The physical process underlying this tunneling spectroscopy is the inelastic scattering of a tunneling electron by a molecule. The energies of the vibrational peaks observed in an IETS spectrum are, to a first approximation, independent of the details of the electron-molecule interaction. Peak shapes and intensities, however are dependent upon the scattering mechanism.

Soon after the experimental discovery of IETS by JAKLEVIC and LAMBE [1], SCALAPINO and MARCUS [2] formulated a theory to explain the order of magnitude of the peak intensities. In their theory the dipole potential of the molecule and its image were included in the barrier potential and the excess tunneling current was estimated for the infrared active modes. LAMBE and JAKLEVIC [3] subsequently considered the possibility of a Raman type of electron-molecule interaction involving molecular polarizability. Their estimate of inelastic current indicated that the Raman mechanism could contribute significantly to the IETS peak intensities.

Recently KIRTLEY, SCALAPINO and HANSMA [4] have examined the IETS intensity problem in more detail. They formulated the problem using the transfer Hamiltonian formalism, and time dependent perturbation theory and obtained theoretical expression for second derivative of the tunneling current with respect to voltage. In their model the incoming and outgoing electron wavefunctions were taken to be decaying along the direction normal to oxide surface and oscillating parallel to it. The interaction between the tunneling electron and the molecule was assumed to be a Coulomb interaction arising from fictitious "partial charges" located on each atom of the molecule. The 'partial charges' and the amplitude of displacement for the vibrational modes were obtained empirically from experimental data. They also examined the orientational dependence of the peak intensities. Their theory was successful in predicting the order of magnitude of the inelastic current and suggested that molecular vibrational modes with atomic displacements normal to the oxide surface would produce more intense IETS peaks then modes whose displacements are parallel to the surface.

Up to this time no calculations of IETS intensities have been carried out using a realistic model for the molecular electronic wavefunctions and the normal modes of vibration. The objective of our study is to carry out a detailed 'ab initio' calculation of the relative IETS intensities resulting from the interaction of a tunneling electron with the oscillating charge density of a vibrating molecule. In these initial calculations we shall consider only the electronic ground state of the molecule. The effect of the vibration of the molecule is calculated from derivative of the molecular charge density with respect to a normal-mode displacement. We employ a 'rigid-ion' approximation, but include the effect of exchange-correlation and renormalization of the molecular wave functions required because of changes in orbital overlap integrals. Image charges can also be included but are neglected in the work reported here.

2. Electron-Molecule Scattering Hamiltonian

In this section we describe the derivation of an approximate scattering potential for the interaction between a tunneling electron and an isolated molecule on the oxide barrier.

We consider a N+1 electron system and label the electron coordinates by $r,s; \xi_1,x_1; \xi_2,x_2 \cdots \xi_N,x_N \cdots$ where r and the set ξ_j are spatial coordinates and s and x_j are spin coordinates. The wavefunction is a Slater determinant with N+1 orthonormal one electron functions: $\psi, \phi_1, \phi_2 \cdots, \phi_N$ where ψ describes the tunneling electron state and the ϕ_j are molecular orbitals. Hartree-Fock equations for ψ are:

$$\left\{ \frac{-\hbar^2}{2m} \nabla^2 + V_s(\vec{r}) - \sum_R \frac{e^2}{|\vec{r}-\vec{R}|} + e^2 \sum_{j=1}^{N} \int d\vec{\xi} \frac{\phi_j^*(\vec{\xi})\phi_j(\vec{\xi})}{|\vec{r}-\vec{\xi}|} \right\} \psi(\vec{r})$$

$$- e^2 \sum_{j=1}^{N} \int d\vec{\xi} \frac{\phi_j^*(\vec{\xi})\phi_j(\vec{r})\psi(\vec{\xi})}{|\vec{r}-\vec{\xi}|} = c\psi(\vec{r}) \qquad (1)$$

In (1), $V_s(\vec{r})$ is the one-electron potential of the solid, consisting of the electrodes and oxide barrier, and $\vec{R}$ represents the positions of the nuclei of the molecule. The terms in the brackets are the kinetic energy, potential of the solid and the Coulomb repulsion between the tunneling electron and the electrons of the molecule. The fourth term on the left hand side of (1) is the exchange potential between molecule electrons and the tunneling electron.

We rewrite (1) in the form:

$$(H^0 + H')\psi(\vec{r}) = \varepsilon\psi(\vec{r}) \tag{2}$$

where:

$$H^0 = \frac{-\hbar^2}{2m}\nabla^2 + V_s(\vec{r}) \tag{3}$$

$$H' = -\sum_{\vec{R}} \frac{e^2}{|\vec{r}-\vec{R}|} + e^2 \sum_{j=1}^{N} \int d\vec{\xi}\, \frac{\rho(\vec{\xi})}{|\vec{r}-\vec{\xi}|} - 6\alpha\left[\frac{3\rho(\vec{r})}{8\pi}\right]^{1/3} \tag{4}$$

In (4), $\rho(\vec{r})$ is the charge density arising from the occupied molecular states $\phi_i = 1,2,\ldots N$ and the last term is the "$X\alpha$" approximation for the exchange potential [5].

A similar equation can be written for each of the molecular states and is of the form:

$$(H_M^0 + H_M') \phi_j(\vec{r}) = E_j\phi_j(\vec{r}) \tag{5}$$

where H_M^0 is the hamiltonian for the isolated molecule. Let $\psi_k^0(\vec{r})$ and $\phi_j^0(\vec{r})$ be the solutions for the isolated systems;

$$H^0\psi_k^0(\vec{r}) = \varepsilon_k^0\psi_k^0(\vec{r}) \tag{6}$$

$$H_M^0\phi_j^0(\vec{r}) = E_j^0\phi_j^0(\vec{r}) \tag{7}$$

If the perturbation potential H' for the tunneling electron is then calculated approximately by using the $\phi_j^0(\vec{r})$ rather than the perturbed molecular states, $\phi_j(\vec{r})$, then H' is given by (4) where now $\rho(\vec{r})$ is the unperturbed molecular charge density.

The perturbation hamiltonian $H' = H'(\vec{r},\{\vec{R}\})$, where $\{\vec{R}\}$ signifies the set on nuclear positions. The dependence of H' on $\{\vec{R}\}$ comes about because of the nuclear Coulomb attraction and the molecular charge density depends on $\{\vec{R}\}$. We make the Born-Oppenheimer approximation and account for the molecular vibrations by writing

$$H'(\vec{r},\{\vec{R}\}) = H'(\vec{r},\{\vec{R}^0\}) + \sum_{\vec{R}_j} \vec{U}_{\vec{R}_j} \cdot \left.\vec{\nabla}H'(\vec{r},\{\vec{R}\})\right|_{\{\vec{R}\} = \{\vec{R}^0\}} \tag{8}$$

where the $\vec{R}_j$ are the elements of $\{\vec{R}\}$ and $\{\vec{R}^0\}$ are the equilibrium positions.

Next we employ first order time dependent perturbation theory to write the rate of transitions between initial and final states as:

$$W_{i \to f} = \frac{2\pi}{\hbar} |<\vec{k}_i|<n_i|H'(\vec{r},\{\vec{R}\})|n_f>|\vec{k}_f>|^2 \delta(E_i-E_f) \tag{9}$$

where E_i and E_f are the initial and final state energies of the molecule-electron system. In (9) $<k_i|$ and $|k_f>$ denote the electron states and $<n_i|$ and $|n_f>$ denote the initial and final molecular vibrational states. In obtaining (9) we have assumed that the electronic state of the molecule is unchanged. The nuclear displacements, U_{R_i}, involved in the second term of (8) can be expanded in the normal modes of vibration of the molecule:

$$\vec{U}_{\vec{R}_i} = \sum_m \left(\frac{\hbar}{2M_i\omega_m}\right)^{1/2} \vec{e}_i(m) \left[b_m^+ + b_m\right] \qquad (10)$$

In (10), ω_m is the angular frequency of the mth normal mode, M_i is the nuclear mass of the i-th atom, and $\vec{e}_i(m)$ is the polarization vector for the displacement of the i-th atom in m^{th} normal mode. b_m^+, b_m are the creation and annihilation operators for a vibration quantum in m^{th} normal mode.

The first term in (8) does not contribute to transitions associated with the molecular vibrations and will be omitted in what follows. The second term of (8) is responsible for the inelastic electron tunneling current arising from the molecular vibrations. From the rate of transitions we may write an expression for the tunneling current as a function of the applied voltage as

$$I(V) = \frac{4\pi e N_S}{\hbar} \sum_{n_i,n_f} P_{ni} \int dk_i \int dk_f |<n_i|<k_i| \sum_{m,j} \left(\frac{\hbar}{2M_j\omega_m}\right)^{1/2}$$

$$\times [\vec{e}_j(m)\cdot\vec{\nabla}_j H'] \times (b_m^+ + b_m)|\vec{k}_f>|n_f|_2^2 f(\varepsilon_f)[1-f(\varepsilon_f-e_V)]$$

$$\times \delta(\varepsilon_i-\varepsilon_f-eV) \qquad (11)$$

In (11), N_S is the number of adsorbed molecules per cm^2; P_{ni} is the thermal probability of the vibrational state $|n_i>$, $f(\varepsilon)$ is the Fermi function, and V is the applied voltage. At T=0, $P_{ni} = 1$, $|n_i> = |0>$, the molecule is in ground state and $f(\varepsilon) = \theta(\mu-\varepsilon)$, where μ is the Fermi energy. Then we have:

$$\frac{d^2I}{dV^2} = \frac{4\pi e^3 N_S}{\hbar} \int dS_i \int dS_f |<\vec{k}_i,\mu| \sum_{j,m} \left(\frac{\hbar}{2M_i\omega_m}\right)^{1/2} \times [\vec{e}_j(m)\cdot\vec{\nabla}_j H']$$

$$|\vec{k}_f,\mu-eV>|^2\delta(eV-\hbar\omega_m) \qquad (12)$$

In (12), the integrals are evaluated on the Fermi surface, $dS_\alpha = \dfrac{d\Omega}{|\vec{\nabla}_{\vec{k}_\alpha}\varepsilon(\vec{k}_\alpha)|}$

and the gradients are evaluated at $\varepsilon_i = \varepsilon_f = \mu$. From (12) we see that the intensity of a peak will depend on $\vec{e}_j(m) \cdot \vec{\nabla}_j H'(\vec{r},\{\vec{R}^0\})$. To determine this function we must calculate the normal modes of vibration of the molecule and the gradient of the molecular charge density. In the next section we discuss the procedure followed to obtain the electronic and vibrational properties of the molecule.

3. Electronic Structure of Ethylene

We have used LCAO (linear combinations of atomic orbitals) method to determine the electronic properties of ethylene. Even though use of LCAO method in Hartree-Fock calculation for molecules is quite common, the use of statistical exchange (Xα) together with LCAO procedure is a recent development [5]. We shall briefly outline the procedure. The effective one particle equation to be solved for the molecule is given by (7) where

$$H_M^0 = -\nabla^2 + V(\vec{r}) \qquad (13)$$

The first term of H_M^0 is the kinetic energy term (in atomic units). The second term, V, consists of the Coulomb potential arising from the nuclear charges and the electron-electron repulsions. There is also another contribution to the potential V arising from the exchange-correlation. The latter is included as prescribed by SLATER and is proportional to $\rho^{1/3}(\vec{r})$ where ρ is the local charge density of the molecule. Thus

$$V = - \sum_j \frac{2Z_j}{|\vec{r}-\vec{R}_j|} + 2 \int \frac{\rho(\vec{r}')d\vec{r}'}{|\vec{r}-\vec{r}'|} + (-6\alpha) \left[\frac{3\rho(\vec{r})}{8\pi} \right]^{1/3} \tag{14}$$

To construct the initial potential in a self-consistent procedure one starts out by expressing $\rho(\vec{r})$ as a sum of atomic charge densities.

$$\rho(\vec{r}) = \sum_j \rho_a^j(\vec{r}-\vec{R}_j) \tag{15}$$

where $\rho_a^j(\vec{r}-\vec{R}_j)$ denotes the atomic charge density arising from the jth atom of the molecule. V can then be expressed as

$$V(\vec{r}) = \sum_j v_a^j(\vec{r}-\vec{R}_j) + \Delta V_{ex}(\vec{r}) \tag{16}$$

where v_a^j, the atomic like potential, is given by,

$$v_a^j(\vec{r}) = - \frac{2Z}{r} + 2 \int \frac{\rho_a(r')d\vec{r}'}{|\vec{r}-\vec{r}'|} + (-6\alpha) \left[\frac{\rho_a(r)}{8\pi} \right]^{1/3} \tag{17}$$

The term $\Delta V_{ex}(\vec{r})$ is the correction term which arises because of expressing $\rho^{1/3}$ in terms of $\rho_a^{1/3}$. This correction is important where the overlap of charge densities from different atoms is significant; that is, where covalent bonds exist.

The molecular orbitals ϕ_m are expanded in terms of the atomic orbitals:

$$\phi_m(\vec{r}) = \sum_{\alpha,i} a_{mi}^\alpha U_i(\vec{r}-\vec{R}_\alpha) \tag{18}$$

where $U_i(\vec{r}-\vec{R}_\alpha)$ denotes the atomic orbital of symmetry type 'i' located at site $\vec{R}_\alpha$. With this form for the molecular wave function, the secular equation which determines the mixing coefficients, a_{mi}^α, and the eigenvalues is given

$$\text{Det} \left\| H_{ij}^{\alpha\beta} - E \, S_{ij}^{\alpha\beta} \right\| = 0 \quad , \tag{19}$$

where

$$H_{ij}^{\alpha\beta} = \int U_i^*(\vec{r}-\vec{R}_\alpha) \, H \, U_j(\vec{r}-\vec{R}_\beta)d\vec{r} \quad , \tag{20}$$

$$S_{ij}^{\alpha\beta} = \int U_i^*(\vec{r}-\vec{R}_\alpha) \, U_j(\vec{r}-\vec{R}_\beta)d\vec{r} \quad . \tag{21}$$

To proceed further the forms of the atomic wavefunction $U_i(\vec{r}-\vec{R}_\alpha)$ and the atomic potential $v_a^j(\vec{r})$ need to be specified. For computational convenience one uses a Gaussian representation for the atomic functions. Thus a typical atomic function may be expressed in the form

$$U(\vec{r}-\vec{R}_\alpha) = \sum_{\substack{j\equiv(\ell,m,n)}}^{M} \sum_{i=1} C_i N_i (\vec{r}-\vec{R}_\alpha)_x^\ell (\vec{r}-\vec{R}_\alpha)_y^m (\vec{r}-\vec{R}_\alpha)_z^n \, e^{-\alpha_i |\vec{r}-\vec{R}_\alpha|^2} . \tag{22}$$

The 'atomic' potential can be fitted to the form

$$v_a^j(|\vec{r}-\vec{R}_\alpha|) = \frac{2\tilde{Z}_j e^{-\alpha_1 |\vec{r}-\vec{R}_\alpha|^2}}{|\vec{r}-\vec{R}_\alpha|} + \sum_{i=2}^{N} \zeta_i e^{-\alpha_j^j |\vec{r}-\vec{R}_\alpha|^2} \tag{23}$$

The parameters $\tilde{Z}_j$, α_i, ζ_i,.... are determined by using a non-linear least square fitting procedure. With these forms for the basis functions and the potential, the integrals occuring in the Hamiltonian matrix elements can be evaluated analytically. The correction term, $\Delta V_{ex}(\vec{r})$, can also be expressed as a linear combination of Gaussians, centered about sites where the over-lapping charge density occurs. Having obtained the Hamiltonian and overlap matrix elements, the secular equation is solved to obtain energy eigenvalues and eigenfunctions.

In these calculations we have not attempted to obtain highly accurate results, however, the one electron energies obtained are in reasonable agreement with ab-initio Hartree-Fock result [7] for ethylene as can be seen from Table 1.

Table 1   One electron energy values (in eV) obtained for ethylene using LCAO-$X\alpha$

Electronic State	EXPT	LCAO-$X\alpha$ ($\alpha = \cdot 90512$)	H-F*
A_{1g}	-23.5	-25.69	-27.5
B_{3u}	-19.1	-18.74	-21.3
B_{2u}	-15.87	-16.86	-17.5
A_{1g}	-14.66	-15.42	-15.3
B_{1g}	-12.85	-13.61	-14.1
B_{1u}	-10.51	-12.29	-10.0

* Reference 7

To obtain the molecular vibrational frequencies and eigenvectors we have employed the method of GWINN [8], and our calculated frequencies are in good agreement with experimental results [9].

Using the calculated vibrational frequencies and eigenvectors and the molecular charge density we proceed to the calculation of the functions, $\vec{e}_\lambda(m)\cdot\vec{\nabla}_\lambda H'$. We note these functions require the calculations of $\vec{\nabla}_\lambda \rho(\vec{r})$. To calculate this quantity we make the "rigid ion" approximation in which the atomic orbitals move with their respective nuclei. This neglects the change in the admixture of orbitals that results when the atoms are displaced. One effect of the change in the mixing coefficients is to maintain the proper normalization of the molecular wavefunctions. This effect can, however, be included by normalization of the displaced rigid ion molecular wavefunctions.

If the atoms of the molecule are displaced by $\Delta\vec{R}_1$, $\Delta\vec{R}_2...\Delta\vec{R}_M \equiv \{\Delta\vec{R}\}$, then to first order in $\{\Delta\vec{R}\}$ the normalized rigid ion wavefunction is

$$\psi_m(\vec{r}, \{\Delta\vec{R}\}) = \phi_m(\vec{r})[1 - \sum_\mu \vec{g}_m^{\,\mu} \cdot \Delta\vec{R}_\mu] + \sum_{i,\mu} a_{mi}^\mu \vec{\nabla}_\mu U_i(\vec{r}-\vec{R}_\mu)\cdot\Delta\vec{R}_\mu \quad , \qquad (24)$$

where

$$\vec{g}_m^{\,\mu} = \sum_{i,j,\lambda} a_{mi}^{*\lambda} a_{mj}^\mu \; U_i^*(\vec{r}-\vec{R}_\lambda)\vec{\nabla}_\mu U_j(\vec{r}-\vec{R}_\mu)d\vec{r} \quad . \qquad (25)$$

The change of the charge density due to the displacements is then

$$\Delta\rho(\vec{r}) = \sum_\lambda \Delta\rho_\lambda(\vec{r}) = 4 \sum_{\lambda,m} \phi_m^*(\vec{r}) \left[\vec{\nabla}_\lambda - \vec{g}_m^{\,\lambda} \right] \phi_m \cdot \Delta\vec{R}_\lambda \; .... \qquad (26)$$

Because $\Delta\rho(\vec{r})$ is a linear sum of contributions due to each displacement separately, the contributions can be calculated independently and then super-imposed to find $\Delta\rho(\vec{r})$ corresponding to any vibrational eigenvector. The con-tribution to the electron-molecule interaction arising from the displacement of the λ-th atom by $\Delta\vec{R}_\lambda$ is:

$$\vec{e}_\lambda(m)\cdot\vec{\nabla}_\lambda H' = - \frac{2Z_\lambda \vec{e}_\lambda(m)\cdot(\vec{r}-\vec{R}_\lambda)}{|\vec{r}-\vec{R}_\lambda|^3} + 8 \sum_{k_{occ}} \sum_{i,j} a_{ki}^{*\mu} a_{kj}^\lambda \; \times$$

$$\left[\int \frac{d\vec{r}'}{|\vec{r}-\vec{r}'|} \left\{ U_i^*(\vec{r}'-\vec{R}_\mu)\vec{\nabla}_\lambda U_j(\vec{r}'-\vec{R}_\lambda) - U_i^*(\vec{r}'-\vec{R}_\mu)U_j(\vec{r}'-\vec{R}_\lambda)\vec{g}_k^\lambda \right\} \cdot \vec{e}_\lambda(m) \right.$$

$$\qquad\qquad (27)$$

$$+ \frac{4c}{\rho^{2/3}(\vec{r})} \sum_{k_{occ}} \sum_{i,j} \left[a_{ki}^{*\mu} a_{ki}^\lambda U_i(\vec{r}-\vec{R}_\mu)\vec{\nabla}_\lambda U_j(\vec{r}-\vec{R}_\lambda)-\rho(\vec{r})\vec{g}_k^\lambda \right]\cdot\vec{e}_\lambda(\vec{m})$$

where $\quad \rho(\vec{r}) = 2 \sum_{k_{occ}} |\phi_k(\vec{r})|^2$

and $\quad c = (-2\alpha) \times \left(\frac{3}{8\pi}\right)^{1/3}$

4. Preliminary Results and Conclusions

Having solved the electronic structure and vibrational problems for the ethylene molecule, we are in possession of $\{a_{kj}^\lambda\}$, $\{\vec{e}_\lambda(m)\}$ and ω_m. Thus we can construct the interaction potential $\frac{\hbar}{2M_\lambda\omega_m}^{1/2} \vec{e}_\lambda(m)\cdot\vec{\nabla}_\lambda H'$, for various modes of vibration.

We have completed calculations of the potential for several types of atomic displacements that characterize the vibrations of ethylene. In this section we present and discuss the results for a typical vibrational mode of ethylene; namely the B_{1g} mode at 3098 cm^{-1}.

Referring to (27), it is seen that there are contributions from three terms. The first is the dipole potential resulting from the vibratory motion of the nuclei. The second is due to the change in the Coulomb interaction between the tunneling electron and electrons of the molecule and the last term is due to the change in the exchange potential resulting from an atomic displacement. The nuclear dipole term requires no discussion and therefore we concentrate on the Coulomb and exchange contributions to the interaction potential.

The Coulomb contribution associated with the displacement of the carbon atom in the B_{1g} vibrational mode of ethylene is shown in Fig. 1. (A schematic of the vibrational displacements is shown at the top of the figure.) It is seen that the potential contours are dipole-like.

Detailed examination of the field indicates that even though its angular dependence is approximately described by cos θ, the radial dependence is more complex. The field is truly dipole-like at distances greater than 4 a.u. from the nucleus. Since the contribution from the nuclear part is purely dipolar in character, the total field which also includes exchange and Coulomb contribution is not dipolar at all points in space. Therefore, the concept of an oscillating partial charge is not valid in detail. In addition the strength of the potential depends upon the displacement direction. For example, the strength for displacement perpendicular to the carbon-carbon bond is found to be approximately 20% greater than that for displacement parallel to the bond.

The situation for the displacement of a hydrogen atom is much more complex as illustrated in Figs. 3 and 4. The Coulomb contribution is shown in Fig. 3. The potential is clearly not dipole-like. Figure 4 shows the sum of the Coulomb and exchange contribution to the interaction potential. The field is very complex and bears no resemblence to a dipolar field. The reason that the hydrogen potential is so complex is that the bond charge associated with the C-H bond does not follow the displacement of the hydrogen nucleus in any simple way.

On the other hand, the dipolar contribution arising from the displacement of the hydrogen nucleus dominates the Coulomb and exchange contributions so that the total potential is still dipole-like.

On the basis of these studies we conclude that the electron-molecule interaction is only crudely represented by a superposition of dipoles located on the vibrating atoms and that the strengths of these dipoles are different for each vibrational mode. This means that different partial charges must be assigned for different modes and therefore the partial charge model cannot be used in a predictive manner.

The next step in this program is to calculate the matrix elements of the interaction potential between initial and final tunneling states. Then, using (11), the intensities of the IETS peaks can be calculated. We do not expect to obtain accurate values for absolute numerical values of the peak intensities, but we expect that the relative intensities should be correct. It is extremely important at this stage of the development of IETS theory to determine whether or not the relative intensities can be qualitatively predicted from calculations such as the one described here. If, in fact, this turns out to be the case then it will be possible to extract significant physical information from IETS spectra. The matrix elements of the interaction depend strongly on the orientation of the molecule relative to the oxide surface. Therefore with a qualitatively predictive model it should be possible to obtain information about molecular orientation directly from IETS peak intensities.

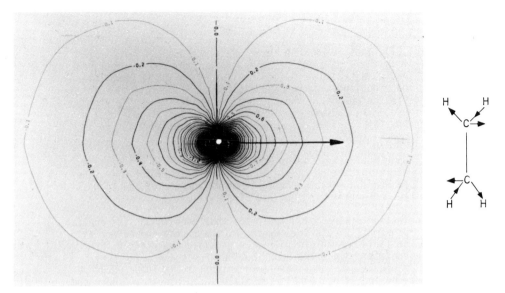

Fig.1 Contours for the Coulomb contribution to the interaction potential for the carbon motion in the B_{1g} mode. Schematics of atomic motions are shown in the insert

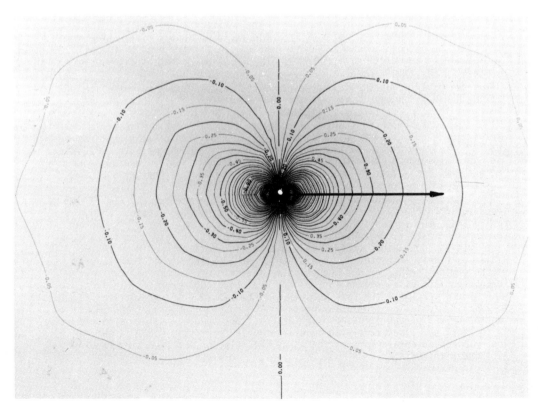

Fig.2 Contours for the Coulomb and exchange contribution to the interaction potential for the carbon motion (see insert in Fig.1) in the B_{1g} mode

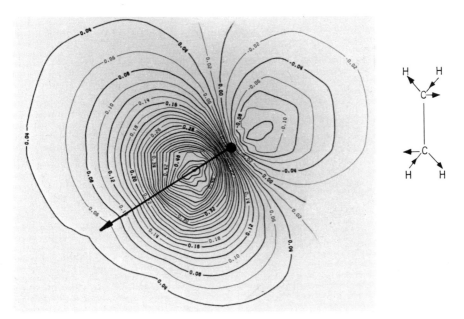

Fig.3 Contours for the Coulomb contribution to the interaction potential for hydrogen motion in the B_{1g} mode. Schematics of atomic motions are shown in the insert

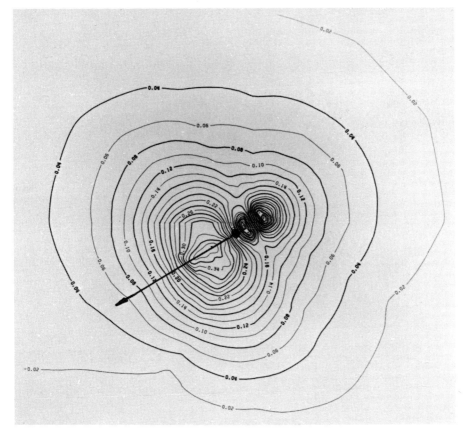

Fig.4 Contours for the Coulomb and exchange contribution to the interaction potential for the hydrogen motion (see insert in Fig.3) in the B_{1g} mode

Acknowledgement is made to the Donors of the Petroleum Research Fund, administered by the American Chemical Society, for partial support of this research.

References

1 R.C. Jaklevic, J. Lambe: Phys. Rev. Lett. 17, 1139 (1966)
2 D.J. Scalapino, S.M. Marcus: Phys. Rev. Lett. 18, 159 (1967)
3 J. Lambe, R.C. Jaklevic: Phys. Rev. 165, 821 (1958)
4 J. Kirtley, D.J. Scalapino, P.K. Hansma: Phys. Rev. B14, 3177 (1976)
5 J.W.D. Connolly in Semi empirical Methods of Electronic Structure Calculation; Part A, ed. G.A. Segal (Plenum Press, New York 1977) pp. 105-132
6 H. Sambe, R.H. Felton: J. Chem. Phys. 61, 3862 (1974)
7 U. Kaldor, I. Shavitt: J. Chem. Phys. 48, 191 (1968)
8 W.D. Gwinn: J. Chem. Phys. 55, 477 (1971)
9 L.M. Sverdlov and others in Vibrational Spectral of Polyatomic Molecule (John Wiley & Sons, 1974) pp. 149-153

Theory of Surface-Plasmon Excitation by Electron Tunneling

L.C. Davis

Research Staff, Ford Motor Company
Dearborn, MI 48121, USA

ABSTRACT

A new calculation of surface-plasmon excitation in tunnel junctions is described. The tunnel junction is divided into three regions of complex dielectric function $\varepsilon_L(\omega)$, $\varepsilon_0(\omega)$, and $\varepsilon_R(\omega)$ which correspond to the left electrode, the barrier, and the right electrode respectively. Maxwell's equations are solved for the classical electromagnetic fields. The source terms are given by the quantum-mechanical transition current and charge,

$$\underline{J} = \frac{ie\hbar}{2m}(\psi_R^\star \nabla \psi_L - \psi_L \nabla \psi_R^\star) \text{ and } \rho = -e\psi_R^\star \psi_L, \text{ for an electron transition from a}$$

state ψ_L in the left electrode to a state ψ_R in the right. The transition rate is given by

$$\frac{-2}{\hbar\omega} \text{Re} \int \underline{E}^\star \cdot \underline{J} d^3r \text{ where } \hbar\omega = E_L - E_R.$$ This new formulation avoids the need

to quantize the electromagnetic fields and allows the use of complex dielectric functions. Numerical estimates of the rate of surface plasmon excitation in $A\ell$-$A\ell_2O_3$-Ag junctions are given.

Surface plasmons are a popular subject now. They show up in a number of experiments. Examples include optical experiments, photoemission, and electron energy loss experiments. Almost any time an electron crosses a boundary between a metal and an insulator it has a good chance of interacting with surface plasmons. My interest in them came about because their importance in tunneling, in particular because of the experiments by JOHN LAMBE and SHAUN McCARTHY at Ford [1]. In these experiments surface plasmons are excited in metal-insulator-metal junctions. The surface plasmons radiatively decay and light from the tunnel junctions is detected. In place of exciting a molecular vibration, the tunneling electrons excite surface plasmons. Of course, there are some tricks to convert them into photons. I am interested in trying to estimate the probability of surface-plasmon emission, particularly which surface plasmons will be emitted because we know that certain ones of them have a good chance to radiate and some of them (that have too high a momenta) cannot radiate. In addition, there are several questions relating to tunneling theory which I want to ask. These relate to what are the electron wave functions, and what is the region of interaction. This goes back to the old question about the transfer Hamiltonian and exact wave functions that Dr. KIRTLEY has already discussed. There are some very simple physical ideas which answer these questions.

Let me briefly mention some of the history of the subject. A few years ago there were two sets of relevant experiments [2]: one by TSUI at Bell Labs and the other by STEINRISSER at the University of Illinois, in which they looked at the I-V characteristics of metal-semiconductor junctions. There was some controversy at the time about what they were seeing but I think eventually everyone agreed that they had seen surface plasmons. They were interested in the second derivative of the current versus voltage. I am not going to be concerned with that additional complication. I am going to look strictly at the rate of excitation of the surface plasmons, because it can be studied by the decay into light. So, we are going to look at a slightly different aspect of inelastic tunneling phenomenon.

At the same time as the experiments were performed a theory was given by NGAI, ECONOMOU, and COHEN [3]. One could modify this theory for metal-in- sulator-metal junctions, but I have chosen not to do that. One of the rea- sons is that I am going to give different answers than they gave to the questions mentioned previously. Secondly, I want to treat the dielectric functions that are involved in a different manner. Now this may be a repeat of some of the things Professor MILLS discussed. We are considering surface plasmons but the ideas are quite similar to surface phonons.

The simplest case one can think of is a metal-vacuum interface. If we neglect retardation (which is permissible for what we are interested in) it becomes a fairly trivial problem. The electric field $\vec{E}$ is simply a gradient of a potential ϕ. The potential has the exponential dependence shown in Fig. 1a, decaying away from the boundary. It is a plane wave in the direc- tion normal to the figure and it is a solution of Laplace's equation $\Delta^2\phi = 0$.

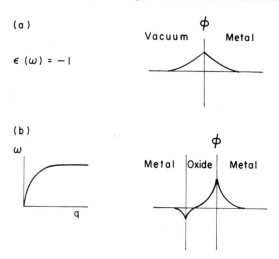

Fig.1 (a) Surface plasmon frequency and potential for a metal-vacuum inter- face. (b) Dispersion curve and potential for a metal-insulator-metal struc- ture

The potential is continuous at the interface as well as the normal component of the displacement. There is a solution when the frequency-dependent di- electric function of the metal, $\varepsilon(\omega)$, is equal to minus one. If one uses a simple free electron dielectric function $\varepsilon(\omega) = 1-\omega_p^2/\omega^2$, where ω_p = the plasmon frequency, one finds that the surface plasmon frequency is $\omega_p/\sqrt{2}$. Now in tunneling we are interested in a sandwich structure. In general the two metals will be different and in this case the potential is shown in Fig. 1b. That is just one of the modes but is the mode we will be interested in. There was a discussion by ECONOMOU [4] several years ago of the various modes of sandwich structures. The dispersion curve is not flat as it is in the metal-vacuum case but, in fact, appears in Fig. 1b. It approaches an asymptote as $q \to \infty$. For the system in which we are interested, $A\ell-A\ell_2O_3-Ag$, this asymptote is at about 3.5 eV. So it is a much different scale of energy than for surface phonons. The region at small ω and q is sometimes called the slow wave region. SWIHART [5] discussed it in connection with super-conducting tunneling.

We require a theory to describe an electron which tunnels inelastically from one side to the other and excites surface plasmon waves. The standard procedure would be to find these modes and to quantize them in some fashion, introducing annihilation operators and the like as was discussed in the previous talk and then put them into a theory found in CHARLIE DUKE's book [6] or NGAI, ECONOMU, and COHEN's paper. However, for several reasons that is not something I wish to do. First of all it is cumbersome to work through all the equations to find the modes and to quantize them. It also becomes a little tricky when the metals themselves have losses in them since the modes may not be well defined. One should use the dielectric functions that have been measured optically which have both real and imaginary parts. I would like to avoid finding modes altogether and it turns out there is an easy way of doing that in this problem (and probably many other problems, too). One does not have to quantize the electromagnetic field at all. Basically it is similar to a calculation of electron energy loss. One starts out with a dielectric system where there are three different dielectric functions: $\varepsilon_L(\omega)$ for the left metal, $\varepsilon_R(\omega)$ for the right and ε_0 for oxide (see Fig.2). They can all be complex and frequency dependent. For the calculation I actually took $\varepsilon_0 = 3$ and did not retain its frequency dependence.

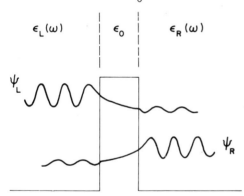

Fig.2 Dielectric functions and wave functions for calculation of inelastic tunneling

When one is considering electron scattering one can often treat the electron classically, but in tunneling this cannot be done. Tunneling is quantum mechanical phenomena so a better treatment of the electron is required. It turns out there is a nice thing that can be done. We can look up in quantum mechanics books like SCHIFF [7] to find the transition charge density J and transition current ρ. (We can use these basic ideas to calculate the radiative decay of the H atom, for example, and it works out just the same as quantizing the fields.) The idea is that an electron starts out in some initial state ψ_L and it makes a transition to a final state ψ_R. That transition sets up a time-dependent current $J = (ie\hbar/2m)(\psi_R^* \nabla \psi_L - \psi_L \nabla \psi_R^*)$ and charge density $\rho = -e\psi_R^* \psi_L$. These are source terms for Maxwell's equations and they generate an electromagnetic field. For example, if the electron starts out with an energy E_L and goes to an energy E_R, the time dependence is $e^{-i\omega t}$ where $\hbar\omega = E_L - E_R$. Now we come to the first question about what wave function should be used for the electrons. We do not have any problem with elastic scattering from the system such as Professor SCHRIEFFER talked about because whatever would be called elastic scattering in this system is all

taken up in the macroscopic dielectric constants. That is all built in to
begin with. The question is should I use the states that BARDEEN [8] first
used in tunneling or should I use eigenfunctions, those Dr. KIRTLEY called
exact solutions? I think the answer is one should use eigenfunctions simply
because the current and the charge density require eigenfunctions. They are
defined in such a way that they satisfy the continuity equation. These other
kinds of functions do not satisfy the continuity equation and one has to insert some surface charges to conserve charge. In fact, I think one of the
conceptual weaknesses of the transfer Hamiltonian approach is that surface
charge is omitted. It probably does not make any difference for inelastic
tunneling from impurities, but it makes a big difference in the surface
plasmon problem.

To simplify things let us talk about electrons which are normally incident
from the left and at the Fermi energy. These are the ones that have the
highest probability of tunneling, so we will focus on their cross section for
interaction with the surface plasmons. As an electron crosses the junction
it is going to transfer some energy $\hbar\omega$ and momentum $\hbar q$ to the electromagnetic
field. Take this momentum to be in the -x direction which is in the plane
of the junction. If we put in the appropriate wave functions then we see
that $\vec{J}$ and ρ have an e^{iqx} and an $e^{-i\omega t}$ time dependence. In the normal direction they are a little more complicated because each wave function con-
sists of an incident wave, a reflected wave, and a transmitted wave. $\vec{J}$ and
ρ become sums of plane waves in the electrodes and exponentials in the barrier but still it is a straightforward problem to work out. Again, if we
neglect retardation it makes the problem even simpler. The electric field
is just $-\vec{\nabla}\phi$ where ϕ is a solution of Poisson's equation, $\nabla^2\phi = -4\pi\rho/\varepsilon$, where
the charge density is given by the transition charge density. The dielectric
function is different for each of the three regions and continuity of the
potential and the normal component of the displacement across the boundary
is required. That is a straightforward but somewhat tedious calculation
which I omit here. The quantity we are interested in is the transition rate.
How does a transition rate come out of such a calculation? We follow elec-
tron loss calculations and look at the energy loss rate $-2\ \mathrm{Re}\int\vec{E}^*\cdot\vec{J}d^3r$. $\vec{J}$
is the transition current density and $\vec{E}$ is the field calculated previously.
This provides the rate at which an electron making a transition from the
initial state to the final state loses energy. The transition rate is just
that energy loss rate divided by the amount of energy lost per transition,
$\hbar\omega$. That is the only concession made to quantum mechanics as far as the
electromagnetic field is concerned. It is treated classically except for
saying the amount of energy has to be $\hbar\omega$. This transition is sort of a
generalization of the golden rule expression that Dr. KIRTLEY wrote down
previously, namely $(2\pi/\hbar)|\langle\psi_L|-e\phi|\psi_R\rangle|^2\ \delta(E_L-E_R-\hbar\omega)$. The broadening of the
delta function due to the losses in the metals is already built into the
new formulation. It is a perfectly good expression for the case where the
modes are not well defined but inelastic tunneling still occurs. So, this
formulation is convenient for real systems where one wants to use the measured dielectric functions. One can even put in a frequency dependent dielectric function for the oxide and look at the excitation of oxide modes.

In the calculations that I am going to show, I want to look at a relative
cross section $r(E_q,\hbar\omega)$ which is the inelastic transition rate divided by the
elastic probability $(E_q \equiv \hbar^2 q^2/2m)$. Now I am hoping by doing that I will
cancel out most of the error made in the treatment of the barrier potential it
would be tedious to treat the potential more accurately. I want to treat it
in the simplest approximation that I can. This transition rate, or shall we

call it a relative cross section, is a function of the amount of energy that is transferred to the electromagnetic field. It is shown in Fig. 3 for a

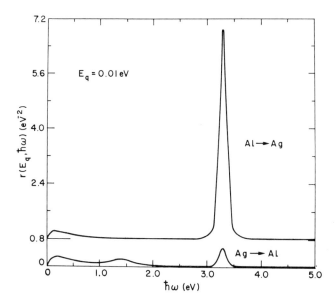

Fig.3 Ratio of the differential probability for inelastic tunneling to the elastic probability vs $\hbar\omega$ for E_q = 0.01 eV (q = 0.051Å^{-1}). The top curve is displaced by 0.8 eV^{-2} for clarity

system of Aℓ-Aℓ$_2$O$_3$-Ag. For fixed momentum transfer, there is a peak at a certain energy. The peak is at the same energy whether the electron goes from Aℓ to Ag or Ag to Aℓ. Now one notices a considerable difference in the intensity of these two peaks. That has to do with asymmetry in tunnel junctions. Let us examine what these modes look like (Fig.1b). For the Aℓ-Aℓ$_2$O$_3$-Ag system, the Ag plasmon frequency is much lower than for Aℓ. The Aℓ frequency is up around 15 eV and the Ag frequency is approximately 4 eV. So in the region of large q, the asymptotic region, the excitation is almost completely confined to the Ag side. That is not true at small q where it is spread throughout the barrier. So one can see that for large energies, 3 to 4 eV, tunneling from Aℓ to Ag is going to be much more probable than going the other way, because the electron can essentially ride the elastic wave almost all the way across the barrier before it has to drop down to the lower energy where it does not tunnel as well. I should make another point with regard to Fig. 1b: the region of interaction between the tunneling electron and the surface plasmon is not just confined to the barrier but includes the electrodes. This is a little different than what happens with the vibrational modes of impurities. Here one has to take into account what happens in the electrodes. If we do not, we will find in some instances a negative energy loss which is clearly unphysical. I think that provides an answer to the second question I asked, what is the region of interaction.

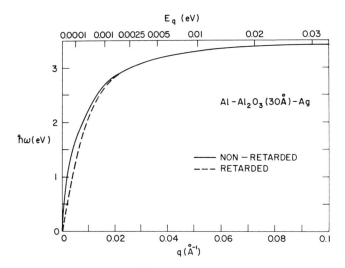

Fig.4 Dispersion curve for surface plasmons in Aℓ-Aℓ$_2$O$_3$-Ag. The oxide thickness is 30 Å

If I follow the peaks in $r(E_q,\hbar\omega)$ as a function of ω and q, I can trace out the dispersion curve shown in Fig. 4. The asymptote is essentially determined by the Ag plasmon frequency. Including retardation does not change the dispersion curve very much. As far as the light emission goes, we can see that any of these modes can be excited if we have enough voltage across the junction. How do they radiate? Here is a real problem. They do not radiate in a nice smooth system, that is both theoretical and experimental fact. All these modes have too much momentum for their energy. So, they have to lose some momentum to couple to the photons. One could consider introducing a grating on one of the surfaces and have the grating soak up some of the momentum. Equally well one could introduce other kinds of random inhomogeneities into the system and they can soak up some of the momentum. In the experiments of LAMBE and McCARTHY, the electrodes are treated in special ways. They are thin but rough and may have pores in them. So, it is difficult to quantify them. All one can do is make some kind of qualitative remarks about which of the surface plasmons, just on a momentum basis, are apt to radiate. Clearly those at small q have less momentum to lose to reach the light line and consequently are more apt to radiate than those that are way out on the asymptote. Now there are many theories about light emission induced by plasmon roughness. Professor MILLS (and others) have written papers on this subject. Basically, if I can make a crude summary, in these theories one finds that the radiative modes are those which have a momentum less than some cut-off determined by the statistical correlation length of the roughness of the surface (or of the film itself). These modes have a higher probability of radiating whereas those with q large compared to this cut-off have no probability of radiating at all. So, it is of interest to look at the relative cross section integrated over the region of the ω-q plane $0 \triangleleft q \triangleleft q_{max}$ where q_{max} is proportional to the reciprocal of the statistical correlation length, and $0 < \hbar\omega < \hbar\omega_{max}$ where $\hbar\omega_{max}$ is just the voltage across the junction. This integrated intensity, $R(E_{max}, \hbar\omega_{max})$ where $E_{max} = \hbar^2 q^2_{max}/2m$, is a crude estimate of the quantum efficiency if we make the assumption that the surface plasmons with $q < q_{max}$ radiate with 100% efficiency and all the others do not radiate at all. For reasonable values of the statistical correlation length (ball-park numbers are known for typical films) the results are shown in Fig. 5. For fixed q_{max}, as $\hbar\omega_{max}$ (voltage) is raised, more and more of the surface plasmons are excited, and depending

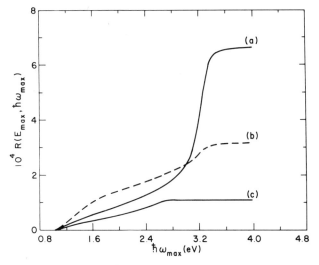

Fig.5 Ratio of inelastic to elastic tunneling probability vs $\hbar\omega_{max}$.

(a) Aℓ to Ag, $E_{max} = \dfrac{\hbar^2 q^2_{max}}{2m} = 0.01$ eV ($q_{max} = 0.051 \overset{\circ}{A}^{-1}$).

(b) Ag to Aℓ, $E_{max} = 0.01$ eV. (c) $E_{max} = 10^{-3}$ eV ($q_{max} = 1.6 \times 10^{-2} \overset{\circ}{A}^{-1}$), no significant difference between Aℓ to Ag and Ag to Aℓ

on how large q_{max} is, various amounts of the asymptote are picked up. There is a large density of states in the asymptote, and that is why the curves in Fig. 5 have a very sharp rise at the asymptotic value. We can see for these typical numbers that the quantum efficiency is of the order of 10^{-4} which is in the ball park of the experiment. It is interesting to see how much energy is really going into the surface plasmons. That is, if one considers the quantum efficiency of emitting a surface plasmon, we find that by letting the cut-off go to infinity we can obtain of the order of 10% quantum efficiency. This means that for every ten electrons which tunnel elastically, one will tunnel inelastically with the emission of a surface plasmon of some energy and some momentum. That is a rather strong interaction and it would be nice if one could convert all of the surface plasmons to photons. To do so would require roughness on an atomic scale, a difficult requirement.

I have plotted the quantum efficiency as a function of E_{max} for the two different bias directions, Ag to Aℓ and Aℓ to Ag, in Fig. 6. One can see

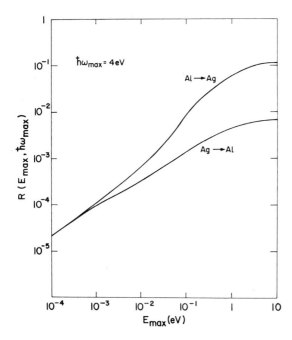

<u>Fig.6</u> Ratio of inelastic to elastic tunneling probability vs E_{max} for $\hbar\omega_{max}$ = 4 eV

that for small E_{max} where the dispersion curve is closer to the light line, and of lower energy, there is not much asymmetry but as we let E_{max} become large there is a big asymmetry between the two directions.

In conclusion, I have presented a simple model of the interaction of electrons with surface plasmons in tunneling. This work has provided an estimate of what the strength is. I also tried to answer in a simple way some fundamental questions in tunneling theory regarding wave functions and the region of interaction. I think my method can be applied to other problems. I think one can probably consider the impurity problem again from the standpoint of treating the dielectric properties of electrodes more carefully, not just putting in an image charge, but putting in the interaction with the electrodes in a manner similar to that described by Professor MILLS.

A more complete version of this work will appear in <u>The Physical Review</u> [9].

References

1 John Lambe, S.L. McCarthy: Phys. Rev. Lett. <u>37</u>, 923 (1976)
2 D.C. Tsui: Phys. Rev. Lett. <u>22</u>, 293 (1969); C.B. Duke, M.J. Rice, F. Steinrisser: Phys. Rev. <u>181</u>, 733 (1969); C.B. Duke: Phys. Rev. <u>186</u>, 588 (1969); D.C. Tsui, A.S. Barker, Jr.: Phys. Rev. <u>186</u>, 590 (1969)

3 K.L. Ngai, E.N. Economou, M.H. Cohen: Phys. Rev. Lett. $\underline{22}$, 1375 (1969);
   K.L. Ngai, E.N. Economou: Phys. Rev. $\underline{B4}$, 2132 (1971)
4 E.N. Economou: Phys. Rev. $\underline{182}$, 539 (1969)
5 J.C. Swihart, Jr.: Appl. Phys. $\underline{32}$, 461 (1961)
6 C.B. Duke: "Tunneling in Solids", (Academic Press, New York, 1966)
7 L.I. Schiff: "Quantum Mechanics",(McGraw-Hill, New York, 1955), 2nd Ed.,
   p. 261
8 J. Bardeen: Phys. Rev. Lett. $\underline{6}$, 57 (1961)
9 L.C. Davis: Phys. Rev. $\underline{B16}$, (1977)

Inelastic Scattering of Low Energy Electron Beams by Surface Vibrations: The Nature of Image Force

D.L. Mills[1]
Department of Physics, University of California
Irvine, CA 92717, USA

ABSTRACT

Experimental studies have examined the inelastic scattering of low energy
electron beams by surface vibrations of crystals with well characterized
surfaces in ultra-high vacuum. We review the data with emphasis on the
systematic features of the electron-surface vibration coupling revealed by
the experiments. This coupling bears an intimate relation to the origin of
the image potential. We also review recent theories of the image potential,
with emphasis on the mechanisms that round off the 1/z divergence obtained
from elementary electrostatics.

1. Introductory Remarks

At this conference, considerable attention has been directed toward the
analysis of inelastic electron tunneling through barriers within which molecules have been selectively introduced. The resolution of this technique is
impressively high, with the consequence that rich spectra may be obtained
for complex molecules.

However, as a number of others have commented also, this technique is quite
limited, in that only a small number of substrates may be employed in these
studies (those which may be incorporated into superconducting tunnel junctions). The substrate surface upon which the molecule is adsorbed is poorly
characterized in the sense of contemporary surface physics, and the presence
of the oxide barrier represents a perturbation whose influence is poorly understood, and difficult to control.

In this paper, I would like to discuss a body of experimental literature
and associated theory complementary to the inelastic electron tunneling work.
This is the inelastic scattering of low energy electron beams from crystal
surfaces prepared under ultrahigh vacuum conditions. Here one uses a highly
mono energetic and well collimated beam of electrons, with the kinetic energy
of the incoming beam of electrons in the range of 5eV-50eV. The basic experiment is quite similar in concept to neutron scattering studies of solids,
except the electron beam samples only the outermost atomic layers of the
crystal, and thus provides information about the vibrational spectrum of the
surface atoms or of molecules and atoms adsorbed on the surface.

The virtue of these inelastic low energy electron diffraction studies
(ILEED) is that one always works with clean and reproducible surface geome-
tries, quite in contrast to those encountered in the inelastic tunneling
experiments. On the other hand, even at best the resolution that can be
achieved in ILEED studies is quite poor (8 meV) by the standard set by tun-

neling spectroscopy. The resolution that can be achieved is quite sufficient for a wide variety of substrates, and substrate/adsorbate combinations to be explored, however. The first high resolution studies of surfaces by ILEED were reported by IBACH [1,2], and subsequently IBACH and his collaborator FROITZHEIM [3] have examined surface vibrations for a variety of systems. A number of laboratories have high resolution ILEED apparatus under construction, so we can expect this to be a rapidly evolving field during the next few years.

The outline of this paper is as follows. In section 2, the basic ILEED experiment will be described, and the nature of the interaction between the electron and the surface vibrations is discussed. As we shall see, the interaction that plays the dominant role in the ILEED spectra reported to date also appears to be the controlling influence in the inelastic tunneling studies. In section 3, we explore what one now appreciates is a closely related topic: a microscopic description of the image potential experienced by a charged particle near the surface of a metal, or a dielectric. Once such a microscopic theory is produced, one may study rounding off of the z^{-1} divergence of the expression for the image potential provided by elementary electrostatics. In section 3, we review the mechanisms that contribute to this rounding, and comment on the relationship between them.

2. Experimental Studies of Inelastic Scattering of Electrons by Surface Vibrations; The Nature of the Coupling

A schematic illustration of the basic ILEED experiment is given in Fig. 1.

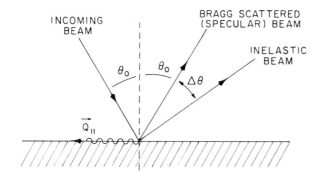

CONSERVATION LAWS: $\vec{k}_{\parallel}^{(s)} = \vec{k}_{\parallel}^{(0)} + \vec{Q}_{\parallel} + \vec{G}_{\parallel}$

$E^{(s)} = E^{(0)} + \hbar\omega_S(\vec{Q}_{\parallel})$

Fig.1 A schematic illustration of an inelastic low energy electron diffraction experiment. In this sketch, the incoming electron emits a surface phonon of wave vector $\vec{Q}_{\parallel}$, to be deflected away from the direction of the Bragg beam. Components of wave vector parallel to the surface are conserved, along with the energy

In the figure, the incoming beam makes an angle θ_0 with the normal, and the electron emits a surface phonon of wave vector $\vec{Q}_\parallel$ upon striking the surface. Wave vector components parallel to the surface are conserved within a reciprocal lattice vector (those normal to the surface are not), and energy is conserved. Thus, the inelastically scattered electron is deflected away from the direction of the Bragg beam by the angle $\Delta\theta$, easily related to $\vec{Q}_\parallel$ and $\hbar\omega_s(\vec{Q}_\parallel)$. Measurement of $\Delta\theta$ and the energy of the outgoing electron allows determination of $\vec{Q}_\parallel$ and $\omega_s(\vec{Q}_\parallel)$. We show here the case of electrons scattered by surface phonons on a clean surface, but the discussion is readily extended to the case where it scatters from a bulk phonon of the substrate, or the localized vibration of an absorbed molecule.

The experimental results and their interpretation are discussed in detail in the review articles cited in footnote [3]. As remarked there, the paper by FROITZHEIM reproduces many of the spectra, and describes the very detailed information one may obtain from them. Here we focus our attention on the nature of the interaction responsible for the scattering. This may be appreciated best by discussing first the case of the ionic crystal ZnO.

The crystal ZnO is an ionic material, very similar to an alkali halide. If we ignore anisotropy, at infrared frequencies the dielectric response of the material is well characterized by the classical oscillator form of the dielectric constant:

$$\varepsilon(\omega) = \varepsilon_\infty + \frac{\Omega_p^2}{\omega^2 - \omega_{TO}^2} \quad , \tag{1}$$

where ω_{TO} is the frequency of the transverse optical phonon of zero wave vector.

A crystal described by the dielectric constant of (1) necessarily has a surface phonon simply described at long wave lengths. We describe this mode. Consider a wave of wave vector $Q_\parallel$ parallel to the x axis, with z normal to the surface. A wave of optical character necessarily sets up an electrostatic potential (we ignore retardation) outside the crystal ($z > 0$) given by $\phi(x,z) = \phi_> \exp(iQ_\parallel x - Q_\parallel z)$, where the z dependence follows from Laplace's equation. Inside the crystal ($z < 0$) for a surface wave, $\phi(x,z) = \phi_< \exp(iQ_\parallel x + Q_\parallel z)$. Conservation of tangential components of $\vec{E}$ across the boundary requires $\phi_> = \phi_<$, while normal components of $\vec{D}$ can be conserved only for a wave with frequency ω_s that satisfies

$$\varepsilon(\omega_s) = -1 \quad . \tag{2}$$

Thus, we have a surface wave with frequency ω_s independent of $Q_\parallel$ [4] given by, with ε_s the static dielectric constant,

$$\omega_s = \omega_{TO} \left(\frac{\varepsilon_s + 1}{\varepsilon_\infty + 1} \right)^{\frac{1}{2}} \quad . \tag{3}$$

Note that ω_s may be calculated from the right hand side of (3) with knowledge only of the macroscopic properties of the crystal.

This mode, a surface mode found on ionic crystals at infrared frequencies, is quite identical to the surface plasmon associated with the metal/vacuum interface. Note that with ω_{TO} set to zero, (1) is the dielectric constant of a nearly free electron metal, if we choose Ω_p^2 to be the plasma frequency of the conduction electrons. If we let $\omega_{TO} \to 0$ in (3) we find $\omega_s = \Omega_p(1 + \varepsilon_\infty)^{-\frac{1}{2}}$, or with $\varepsilon_\infty = 1$ as frequently presumed for simple metals, we have the well-known result $\omega_s = \Omega_p/\sqrt{2}$.

With this information in hand, we turn to the ILEED data reported by IBACH [1], for electrons reflected from the clean surface of ZnO.

IBACH finds a sequence of intense energy loss peaks at the energies $n \hbar \omega_s$, with ω_s given by (3) and with n as large as 5. The inelastically scattered electrons associated with these features emerge very near the specular direction. Indeed, the angular distribution cannot be resolved with the apparatus, which has an angular resolution $\Delta \theta \sim 1°$.

In the case of ZnO, the data may be understood quite quantitatively by invoking a simple scattering mechanism. As we saw in the discussion that preceeds (3), a surface phonon of wave vector $Q_\parallel$ sets up an electric field that extends a distance $\ell = Q_\parallel$ into the vacuum above the crystal. For small $Q_\parallel$, ℓ is very large. As the electron approaches the crystal, it may excite a surface phonon by coupling to this long-ranged electric field. Theories which invoke this mechanism either by treating the electron classically [5] or quantum mechanically [6] provide a fully quantitative description of the data.

The essential features of the scattering process may be appreciated by a very simple argument. Let the electron beam strike the surface at normal incidence ($\theta_0 = 0$ in Fig. 1). Then $k_\parallel^{(s)} = Q_\parallel = k_0 \Delta \theta$ from wave vector conservation. In the last step, k_0 is the wave vector of the incident electron, and $\Delta \theta$ is assumed small. We may then write $\Delta \theta = \hbar Q_\parallel / p_0$, with p_0 the momentum of the incoming electron. A phonon of wave vector $Q_\parallel$ is excited most efficiently by an electron that spends a time $\Delta t \approx 2\pi/\omega_s$, the period of the surface wave, under the influence of its electric field. (If $\Delta t << 2\pi/\omega_s$, the adiabatic theorem tells us the surface wave will not be excited, and clearly it will be only weakly excited when $\Delta t >> 2\pi/\omega_s$.) But also $\Delta t \approx 2 Q_\parallel / v_0$, and we thus have $Q_\parallel \approx \omega_s/v_0$ for strongest excitation. Typical values of $\Delta \theta$ are then $\hbar\omega_s/E_0$ with E_0 the incident energy. For ZnO, one has $\Delta \theta \sim 0.1°$.

The above argument shows that the strong small angle scattering reported by IBACH has its origin in the coupling of the electron to oscillating electric fields set up by the surface wave in the vacuum <u>above</u> the crystal.

One might believe that the coupling mechanism outlined above is very special to an ionic crystal like ZnO. We have dwelt on the discussion for some time because precisely the opposite is true. To date, all experimental studies of inelastic electron scattering from vibrations on crystal surfaces show this strong lobe of inelastically scattered electrons centered about the specular (or Bragg) beams. This is true for homopolar materials like silicon [2], and also for scattering from vibrations of adsorbed species on the surface.

In the case of the ionic crystal ZnO, the origin of the electric field is clear, since long wave length lattice disturbances of optical character set up macroscopic electric fields in general. In a homopolar crystal like silicon, a more subtle and highly surface specific mechanism is responsible for the electric fields set up by surface phonons. Since an atom in the surface necessarily sits at a site which lacks inversion symmetry, it has a static dipole moment. As the atom vibrates, the dipole moment is modulated by the atomic motion, and the oscillating dipole sets up an oscillating electric field above the crystal. Laplace's equation in the vacuum assures this electric field has the same spatial variation as in ZnO $(exp(iQ_\| x-Q_\| z))$, although in a crystal such as silicon its source is the outermost atomic layer or two, while in ZnO all of the atoms which participate in the surface mode contribute to the field, even those well in the bulk. The scattering observed from the silicon surface, where a single loss peak is seen from the 2 x 1 reconstructed cleavage surface, is thus much weaker than the loss peaks seen in ZnO. Nonetheless, one still has the narrow angular distribution with $\Delta\theta \sim \hbar\omega_S/E_0$.

In silicon, the loss peak observed on the clean reconstructed 2 x 1 surface is quite sensitive to the details of the surface geometry. The loss peak disappears when the crystal is heated to form the complex 7 x 7 reconstructed pattern present at high temperatures, and it disappears when oxygen is present on the surface, to be replaced by characteristic loss peaks associated with the oxygen. This confirms that the dynamic dipole moments are associated primarily with the outermost atomic layer.

For vibrations of atoms adsorbed on the surface, again the strong small angle scattering is found. The electrons now couple to electric fields set up by the oscillating dipole moment of the adsorbed species.

We thus see that in all of the cases studied so far a simple and universal mechanism controls the scattering intensity. This has allowed quantitative information on the magnitude of the dynamic effective charge to be extracted from the measurement of the intensities of the loss peaks [6,7]. As a test of this procedure, IBACH has compared the dynamic effective charged for CO on Pt obeyed from electron scattering data to that obtained from infrared studies, to find the two are in remarkable quantitative agreement [7].

A coupling mechanism identical to that described above has been invoked to explain the left/right asymmetries found in inelastic electron tunneling spectroscopy, as well as the magnitude of the inelastic tunneling structures [8]. It is intriguing to see that these two experiments, very similar in underlying principal but very different in execution, are controlled by the simple electric dipole coupling.

The present discussion of the ILEED experiments have focused entirely on the intense lobe at small angles controlled by electric dipole scattering, since the experimental studies reported to date examine only this feature. There is also a significant intensity of inelastics (produced by scattering from phonons) that emerge at angles large compared to the characteristic angle $\Delta\theta = \hbar\omega_S/E_0$ that entered our discussion of electric dipole scattering. These electrons are scattered not by long ranged electrostatic fields, but by interacting directly with the vibrating ion cores via the electron-phonon interaction. The presence of these electrons is evident in all LEED experiments; they form the thermal diffuse background that becomes more intense with increasing temperature. While the angular dependence of the thermal diffuse background has been studied, the energy distribution of the electrons scattered through large angles has yet to be studied experimentally. A theoretical analysis of a simple model [9] shows that such spectra provide detailed information about the lattice dynamics of the crystal surface hard to obtain by other means. The development of electron spectrometers capable of study of the energy spectrum of the diffusely scattered electrons will thus be a major advance in surface physics.

The discussion of the present section has been confined to the inelastic scattering of low energy electron beams by the vibrational motion of surface atoms, or that of adsorbed species. A number of experimental studies explore the elastic scattering of low energy electrons by collective excitations of electrons in metals [10] (surface plasmons, volume plasmons), and recently loss processes have been studied where the incoming electron excites an electron in the valence band of a semiconductor across the band gap into the conduction band [11], or from a deep core level into the conduction band [12].

The data on the small angle scattering produced by electronic excitations across the band gap of semiconductors can be interpreted quantitatively [11] by invoking precisely the same mechanism described above in the discussion of scattering from surface vibrations. The electron-hole excitation can be regarded as a charge fluctuation of electric dipole character that sets up a macroscopic electric field outside the crystal, in precisely the same fashion as the surface phonon in zinc oxide [13]. One may relate the electron energy loss cross section from bulk processes to the frequency dependent dielectric cross section of the substrate, to obtain a relation that provides a good quantitative account of the data on silicon and germanium [3,11].

3. The Theory of the Image Force on a Charged Particle

In this section, we turn our attention to a topic that at first glance appears to be distinctly different than that of the preceding section. This is the origin of the image force on a charged particle outside (or inside) the surface of a dielectric or metallic crystal, examined from a microscopic point of view. We saw in section 2 that the small angle inelastic scattering of electrons from either surface vibrations of the crystal or from electronic excitations is controlled by coupling of the incoming electron to electric fields set up outside the crystal by the electric dipole moment associated with the excitation. It is this same interaction that gives rise to the image force on an electron near the surface. To describe the image force or image potential, we must consider the self-energy of the charged particle near the surface, where up to now it is loss processes rather than energy shifts that have been of concern to us.

The origin of the image potential, or image force, is best appreciated through a simple example. Consider an electron above an ionic crystal with the dielectric constant of (1), but with the background dielectric constant ε_∞ set equal to unity. Thus, we ignore the electronic excitations, and the dielectric constant of the material differs from unity only by virtue of the electric dipole moment associated with the long wavelength transverse optical phonon. As remarked after (3), by letting $\omega_{TO} \to o$ in all the results, one obtains results appropriate to an electron near the surface of a semi-infinite metal, where interband contributions to the dielectric constant have been ignored.

The model just described has been explored in detail by EVANS and MILLS [14], for the case where the electron is inside as well as outside the model crystal. We summarize their results here, and confine our attention to the case where the electron is outside the crystal. Then the electron couples to the surface phonon with frequency ω_s given in (3), and the Hamiltonian may be written

$$H = \frac{p^2}{2m} + \hbar \, \omega_s \sum_{\vec{Q}_{\parallel}} \alpha^+_{Q_{\parallel}} \alpha_{Q_{\parallel}} + V_s \quad , \tag{4}$$

where the first term is the kinetic energy of the electron, the second the excitation energy of the surface waves, and V_s the interaction between the electron and the surface wave. From Section 2, clearly V_s has the form

$$V_s = \frac{1}{A^{\frac{1}{2}}} \sum_{\vec{Q}_{\parallel}} \Gamma_s(Q_{\parallel}) \left(\alpha_{Q_{\parallel}} + \alpha^+_{-Q_{\parallel}} \right) \exp \left(i \, \vec{Q}_{\parallel} \cdot \vec{r}_{\parallel} - Q_{\parallel} z \right) \tag{5}$$

where A is the surface area. A simple argument shows that

$$\Gamma_s(Q_{\parallel}) = 2\pi e^* \left(\frac{n\hbar}{2M\omega_s Q_{\parallel}} \right)^{\frac{1}{2}} \quad , \tag{6}$$

where e^* is the transverse effective charge of the unit cell, M its reduced mass, and n the density of unit cells in the crystal.

The Hamiltonian of (1) has the same form as the Hamiltonian that describes polaron motion in ionic crystals, and may be studied by methods very similar to those used in polaron theory [14,15], for the case where the electron moves slowly near the surface.

Let us first consider a simple and trivial limit of the Hamiltonian in (4). Suppose the kinetic energy of the electron can be ignored. Formally, we imagine the mass of the electron to be infinite, so it remains fixed at position z. If the electron kinetic energy is ignored, then as MAHAN notes [16], the Hamiltonian is trivially diagonalized by transforming to displaced oscillator variables $\alpha_{Q_{\parallel}}$ given by

$$\alpha_{Q_{\parallel}} = \alpha_{Q_{\parallel}} + \frac{\Gamma_s(Q_{\parallel})}{A^{\frac{1}{2}}} \exp(i\vec{Q}_{\parallel} \cdot \vec{r}_{\parallel} - Q_{\parallel} z) \quad . \tag{7}$$

In terms of the new variables, the Hamiltonian becomes

$$H = \hbar\omega_s \sum_{\vec{Q}_{\parallel}} \alpha^+_{Q_{\parallel}} \alpha_{Q_{\parallel}} - \frac{1}{A} \sum_{\vec{Q}_{\parallel}} \frac{\Gamma^2_s(Q_{\parallel})}{\hbar\omega_s} \exp(-2Q_{\parallel} z) \tag{8a}$$

$$= \hbar\omega_s \sum_{\vec{Q}_{\parallel}} \alpha^+_{Q_{\parallel}} \alpha_{Q_{\parallel}} - \frac{e^2}{ez} \left(\frac{\varepsilon_s - 1}{\varepsilon_s + 1} \right) \quad , \tag{8b}$$

where (8b) follows upon evaluating the sum on $\vec{Q}_{\parallel}$ in (8a), and introducing

118

the static dielectric constant $\varepsilon_S = 1 + \Omega_p^2/\omega_{TO}^2$ of model crystal.

The second term in (8b) is the image potential experienced by a particle near the surface. When a charge is placed near the crystal surface, the coupling between the charge and the excitations in the crystal (in this case the coupling between the charge and the surface excitation) produces a static polarization that sets up an electric field which reacts back on the charge, and lowers its energy. To obtain the (repulsive) image force experienced by a charge <u>inside</u> the crystal, it is necessary to supplement the Hamiltonian of (4) by adding the coupling between the electron and the longitudinal optical phonons (or the bulk plasmons, for an electron in a metal) [14].

On physical grounds, it is clear that the z^{-1} dependence of elementary electrostatic theory cannot hold for very small values of z. The image potential must "round off" in some fashion. Now that we have a microscopic theory of the image potential, we can explore the way this happens.

If we proceed with the Hamiltonian of (4), we must repeat the argument that leads to (8b), but the role of the electron kinetic energy term $p^2/2m$ must be incorporated into the discussion. To do this properly leads one into the mathematical structure of polaron theory [14,15]. However, the principal qualitative features may be appreciated from a simple argument.

The second term of (8a) has the form of the self-energy of a massive particle calculated in second order of perturbation theory. If an infinitely massive particle is placed at z outside the crystal, its self energy is lowered because it can emit and readsorb surface phonons. The matrix element for emission is $\Gamma_S(Q)A^{-\frac{1}{2}}\exp(i\vec{Q}\cdot\vec{r} -Q z)$, that for absorption is $\Gamma_S(Q)A^{-\frac{1}{2}}\exp(-i\vec{Q}\cdot\vec{r} -Q z)$, and the energy denominator is $\hbar\omega_S$, since a surface phonon appears in the intermediate state.

The above remarks apply to a particle of infinite mass. If the mass is finite, the particle recoils upon emitting the surface phonon. If the initial kinetic energy of the particle is small, then since wave vector is conserved in the interaction with the surface phonon, in the intermediate state the particle has kinetic energy $\hbar^2 Q_\parallel^2/2m$. We must then correct the energy denominator of the second term in (8a) by adding the recoil energy to the denominator. This gives an effective image potential $V_{EFF}(z)$ of the form

$$V_{EFF}(z) = -\frac{1}{A} \sum_{\vec{Q}_\parallel} \Gamma_S^2(Q_\parallel) \frac{\exp(-2Q_\parallel z)}{\hbar\omega_S + \frac{\hbar^2 Q_\parallel^2}{2m}} \, . \tag{9}$$

The behavior of $V_{EFF}(z)$ is sketched in Fig. 2. For values of $z > z_S$, where $z_S = (\hbar/2m\omega_S)^{\frac{1}{2}}$, $V_{EFF}(z)$ reduces to the image potential of elementary electrostatic theory. For $z < z_S$, there are deviations from the classical form, and most importantly $V_{EFF}(z)$ assumes the finite value indicated in Fig. 2 at z = 0.

The above discussion applies to slowly moving electrons, i.e. those for which the recoil energy in the denominator of (9) is large compared to the initial kinetic energy. This condition will be satisfied for electrons with initial kinetic energy small compared to $\hbar\omega_S$, for practical purposes.

Also, the expression in (9) provides only a crude representation of the image potential in the regime $z \leq z_s$. The detailed theory [14] leads one to a Schrödinger equation with a non-local potential. This equation must be solved in a self consistent fashion. Nonetheless, the recoil effects discussed above "round off" the effective image potential in a manner qualitatively consistent with (9), for $z < z_s$.

For optical phonon frequencies typical of semiconductors, one finds $z_s \cong 10$ Å, so for slow electrons, the classical description of the image potential fails for $z \leq z_s$.

Our model contains only the lattice contribution to the dielectric constant. There is always an electronic contribution in real materials, and the parameter ε_∞ of (1) differs substantially from unity. The same discussion may be applied to the contribution to the image potential from coupling to the electronic excitations. The electronic contribution retains the classical form down to the much smaller distances $(\hbar/2mE_g)^{1/2}$, where E_g is the band gap.

For electrons near the surface of a simple nearly free electron metal, as we have seen, we can apply the present formulae by taking the limit $\omega_{TO}^2 \to 0$, interpreting Ω_p^2 as the electron plasma frequency, and replacing ω_s by the surface plasmon frequency $\Omega_p/1 + \varepsilon_\infty)^{1/2}$. Then for metals $z_s \sim 1$Å, and the classical form of the image potential holds until we get very close to the surface.

When the particle is within one Angstrom of the surface of a metal other effects not included above play an important role. One of these may be appreciated by returning once again to the limit of infinite mass, so the recoil effect may be ignored. As the charged particle is brought very near the surface, the radius of the static screening charge induced by it becomes progressively smaller; the radius is approximately equal to the distance of the change from the surface. In a metal, the radius of the screening charge can become no smaller than the Fermi-Thomas screening length λ_{FT}. Thus, for $z \leq \lambda_{FT}$, even in the limit of infinite mass, there are deviations from the classical form of the image potential [17]. The form of the image potential for a semi-infinite nearly free electron metal has been discussed by NEWNS [18], and by LANG and KOHN [18], for the infinite mass limit.

For electrons near the surface of a semi-infinite, nearly free electron metal, the ratio $\lambda_{FT}/z_s = (\hbar\omega_p/2\sqrt{2}\ E_F)^{1/2}$, where ω_p and E_F are the plasma frequency and Fermi energy, respectively. This ratio is near unity for metals, so a description of the image potential for slow electrons near metal surfaces should include the recoil effect, in addition to limiting the size of the induced polarization cloud. This is a complex problem yet to be addressed in the theoretical literature. For an ion of mass M near the surface, the ratio λ_{FT}/z_s is reduced by the factor $(\overline{m/M})^{1/2}$, and the recoil effect is clearly unimportant.

The above remarks discuss in qualitative terms the principal effects which lead to rounding off of the image potential, for a slowly moving charged particle near the surface of a dielectric or metal. While the quantitative theory is both complex and incomplete at present, the arguments lead one to appreciate the fundamental length scales relevant to the rounding off phenomenon.

The comments so far apply only to slow charged particles, with kinetic energy small compared to the excitation energy $\hbar\omega_s$ of the fundamental surface excitations (surface phonons, plasmons) to which the charge particle couples. In the opposite limit of large kinetic energy, it is possible to approach the problem in classical language, since the correspondence principle insures validity of the classical result. We discuss the effective image force exerted on a rapidly moving charged particle, from the point of view offered recently by the present author [19].

In the limit of large kinetic energy, the motion of the charged particle can be treated entirely in classical terms. We have a charged particle which moves near a dielectric and excites the dielectric through its coulomb field. The induced polarization in the dielectric sets up an electric field that acts back on the particle. The dielectric does not respond instantly to the presence of the moving particle, but there is a time lag in its response. When this time lag is included, one obtains a force on the particle that has the form of the classical image force far from the surface, but deviations occur as the particle approaches the surface.

In this circumstance, it is important to note that we can no longer speak of an image potential in a meaningful manner. For example, in classical physics, potentials depend on the present position of the particle, and are independent of the velocity, and most particularly on the past history of its motion. From the preceding paragraph, quite clearly the force on the

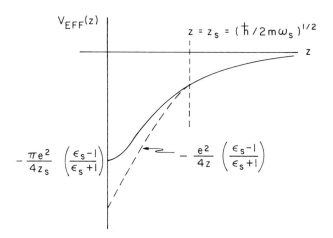

Fig.2 The effective image potential of (9)

particle at any moment depends not only on its present position and its present velocity, but also on the details of the trajectory it has followed in earlier times. Thus, while the force exerted on the particle at any instant is a clear and unambiguous classical concept, this force can in no sense be derived from a potential, in contrast to the implications in a number of recent papers on the topic. In quantum mechanics, one is presumably lead to a self-consistent Schrödinger equation to solve, rather like that one obtained by EVANS and MILLS [14] in their discussion of slowly moving particles near surfaces, although the generalization to rapidly moving particles (kinetic energy greater than $\hbar\omega_S$) has yet to be made.

When a rapidly moving charged particle is far from the crystal surface, the image force assumes the form provided by elementary image force theory, as long as its motion is non-relativistic [19]. For a material with the dielectric constant of (1), there is a "rounding off" of the image force at distances $z_S = v/\omega_S$, where ω_S is the surface wave frequency given in (3). We now are presuming the kinetic energy of the particle is large compared to $\hbar\omega_S$, so notice $v/\omega_S \gg (\hbar/2m\omega_S)^{\frac{1}{2}}$. While there is as yet no complete theory which takes us form the regime of low energies where the healing length is $(\hbar/2m\omega_S)^{\frac{1}{2}}$ to the regime of large kinetic energy ($\gg\hbar\omega_S$) where it is v/ω_S, the velocity dependence of the healing length is as indicated in Fig. 3, with $(\hbar/2m\omega_S)^{\frac{1}{2}}$ a quantum limit achieved at low velocities.

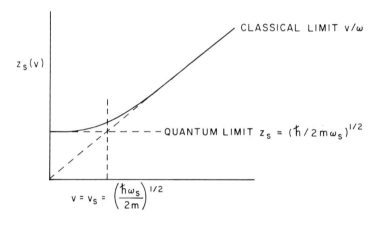

Fig.3 A sketch of the velocity dependent healing length of image potential theory illustrating the passage from the quantum limit relevant to slowly moving particles to the classical limit valid for fast particles

For a nonrelativistic particle which approaches the surface of a dielectric from normal incidence described by the dielectric constant of (1) with ε_∞ frequency dependent, the image force $F_<(z)$ is [19]

$$F_<(z) = \begin{cases} -\dfrac{e^2}{4z^2}\left(\dfrac{\varepsilon_S-1}{\varepsilon_S+1}\right) & , \quad z > z_S \\[3mm] -\dfrac{e^2}{4z^2}\left(\dfrac{\varepsilon_\infty-1}{\varepsilon_\infty+1}\right) - \dfrac{2e^2\Omega_p^2}{v^2(1+\varepsilon_\infty)^2}\,\ln\left(\dfrac{v}{2\omega_S z}\right), & z < z_S \end{cases} \tag{10}$$

where again $z_S = v/\omega_S$ in this result. The contribution to the image force from the electronic excitations that contribute to ε_∞ "round off" at the much shorter distance v/ω_x, where ω_x is a measure of the electronic excitation energy.

A striking illustration of the dependence of the image force on the past history of the particle as well as its present position is found by contrasting the result in (10) valid for a particle which <u>approaches</u> the surface along the normal with that valid for a particle at the same position z which <u>recedes</u> from the surface along the normal after it has passed through the surface. The latter case is of interest in photoemission experiments. On the receding particle, the force is $F_>(z)$ given by [19], for $z > z_s$,

$$F_>(z) = F_<(z) + \Delta F(z) \tag{11}$$

where

$$\Delta F(z) = -\frac{8e^2\Omega_p^2 v}{(1+\varepsilon_\infty)^2 \omega_s^3} \frac{1}{z^3} \sin\left(\frac{z}{z_s}\right) \quad . \tag{12}$$

The physical origin of $\Delta F(z)$ is clear. A particle that passes through the surface excites surface waves coherently (the same surface waves discussed in section 2), and these waves set up macroscopic electric fields which generate the component $\Delta F(z)$ of the force felt by the outgoing particle.

We have outlined the theory of the image potential or image force felt by a charged particle near a surface, with emphasis on the nature of the mechanisms that round off the strong divergence present in elementary electrostatics. We see that one has presently in hand descriptions of this rounding off phenomenon valid in certain limiting regimes, but as yet no unified theory of the phenomenon exists capable of extrapolating from one regime to other, or incorporating all the rounding effects one believes to be important. An example of the former is the absence of a theory which describes both the quantum and classical regimes of Fig. 3, instead of the two very different approaches used in each. An example of the latter is the absence of a theory which incorporates both the rounding from the recoil effect, and the finite Fermi-Thomas screening length for a slow electron near the surface of a metal.

[1]Supported by Grant No. AFOSR of the Air Force Office of Scientific Research, Office of Aerospace Research, U.S.A.F. Technical Report No. 77-50

References

1 H. Ibach: Phys. Rev. Letters <u>24</u>, 1416 (1970)
2 H. Ibach: Phys. Rev. Letters <u>27</u>, 253 (1971)
3 We refer the reader to two review articles for a more complete review of the experiments performed to date, and related theory. See H. Froitzheim, Vol. 4 <u>Topics in Current Physics</u>, edited by H. Ibach, (Springer, New York, 1976). This paper reproduces many of the experimental spectra. The present author has prepared a review article that will appear shortly (D.L. Mills, <u>Progress in Surface Science</u>, to be published)

4   The frequency ω_s is independent of $Q_\parallel$ only when retardation effects are ignored. Retardation effects are unimportant for $cQ_\parallel \gg \omega_s$, where c is the velocity of light. More generally $\omega_s(Q_\parallel) = cQ_\parallel[\varepsilon(\omega)/(1+\varepsilon(\omega))]^{\frac{1}{2}}$, a result equivalent to (4) when $cQ_\parallel \gg \omega_s$. The surface mode frequency $\omega_s(Q_\parallel)$ lies between ω_{TO} and the limiting frequency displayed in (4). The electron scattering experiments probe values of $Q_\parallel$ in the range 10^6 cm^{-1}, and $\omega_s/c \sim 10^3$ cm^{-1}, so for the present discussion neglect of retardation effects is well justified

5   A.A. Lucas, M. Sunjic: Prog. Surf. Sci. 2, 75 (1972)
6   E. Evans, D.L. Mills: Phys. Rev. B5, 4126 (1972) and B7, 853 (1973)
7   H. Ibach (to be published)
8   J. Kirtley, D.J. Scalapino, P.K. Hansma: Phys. Rev. B14, 3177 (1976)
9   V. Roundy, D.L. Mills: Phys. Rev. B5, 1347 (1972)
10 For example, see C.J. Powell: Phys. Rev. 175, 972 (1968)
11 H. Froitzheim, H. Ibach, D.L. Mills: Phys. Rev. B11, 1980 (1975)
12 See R. Ludeke, L. Esaki: Phys. Rev. Letters 32, 653 (1974)
13 A reasonably detailed discussion has been presented by D.L. Mills: Surface Science 48, 59 (1975)
14 E. Evans, D.L. Mills: Phys. Rev. B8, 4004 (1973)
15 T.D. Clark: Solid State Communications 16, 861 (1975)
16 G.D. Mahan in Collective Properties of Physical Systems, edited by B. Lundqvist and S. Lundqvist (Nobel Foundation, Stockholm).
17 A crude description of these deviations follows upon calculating the second term on the right hand side of (8a) not by integrating over all $Q_\parallel$ as in the calculation that leads to (8b), but by cutting off the $Q_\parallel$ integration at the value $Q_\parallel = Q_c$, where Q_c is the inverse of the Fermi-Thomas screening length.
18 D.M. Newns: J. Chem. Phys. 50, 4572 (1969), N.D. Lang, W. Kohn: Phys. Rev. B7, 3541 (1973)
19 D.L. Mills: Phys. Rev. B15, 763 (1976)

Effect of Cooperative Behavior on Molecular Vibrational IETS Peak Intensities [1]

Stephen L. Cunningham and W. Henry Weinberg [2]

Division of Chemistry and Chemical Engineering, California Institute of Technology, Pasadena, CA 91125, USA
and
John R. Hardy

Behlen Laboratory of Physics, University of Nebraska
Lincoln, NB 68508, USA

ABSTRACT

Contrary to previous theories of the intensity of molecular vibrational energy loss peaks in IETS, we have shown that the second derivative of the tunneling current varies as $n^{4/3}$ rather than linearly in n, where n is the concentration of molecules adsorbed on the insulator surface during the junction fabrication. This is a result of considering the proper form for the potential in the junction region from a layer of vibrating dipoles with all of its images. Even though obtained with a simple model for the junction, this dependence agrees with the experimental results of LANGAN and HANSMA.

Recently, LANGAN and HANSMA [1] adsorbed tritiated benzoic acid (C_6H_5COOH) of concentration c in aqueous solution on an aluminum oxide surface. Using a scintillation counter, they determined that the surface concentration n varied an $c^{1.2}$. Separately, upon fabricating an Al-Al_2O_3Pb tunnel junction with benzoic acid adsorbed on the oxide surface at the Al_2O_3-Pb interface, they found that the intensity of an Inelastic Electron Tunneling (IET) feature at 686 cm^{-1} varied as $c^{1.6}$ [1]. Consequently, the IET intensity varies with surface concentration as $n^{1.3\pm0.1}$[2].

The purpose of this paper is to show that a non-linear dependence between intensity and concentration is a consequence of the cooperative behavior of the adsorbate molecules, i.e., the fact that the molecules do not vibrate independently but rather as normal modes of the system. This cooperative behavior manifests itself since the interaction potential between the tunneling electron and the adsorbate layer is not simply n times the potential of a single adsorbate molecule. Rather, the phase relationship between the vibrations of different molecules is important and significantly modifies the potential.

In a previous paper [2], a simple model was used in which the normal modes of the adsorbate layer were plane waves, a consequence of assuming that the adsorbate layers was ordered. Within the dipole approximation, the transition matrix element between a free electron state at the Fermi level E_F on the left side of the barrier and a free electron state at the Fermi level E_F' on the right side of the barrier was shown to be

$$\langle \vec{k}'|U|\vec{k}\rangle = \frac{A\,e^{-\kappa's}k_z k_z'\,\delta(\vec{k}'-\vec{k}-\vec{q})}{\sinh qs}$$
$$\times\left\{\frac{1-e^{-(\Delta+q)s}}{\Delta+q} - \frac{1-e^{-(\Delta-q)s}}{\Delta-q}\right\}, \tag{1}$$

where $\Delta = \kappa - \kappa'$. $\tag{2}$

125

Here, $\vec{k}$ is a two-dimensional wave vector parallel to the barrier labeling the initial state, and

$$k_z = \left(\frac{2mE_F}{\hbar^2} - k^2 \right)^{1/2} \quad , \tag{3}$$

and

$$\kappa = \left(\frac{2m\phi}{\hbar^2} + k^2 \right)^{1/2} \tag{4}$$

Quantities with a prime are defined similarlyfor the final state. The barrier width is s, and the barrier height is ϕ. The wave vector $q = |\vec{q}|$ describes the plane wave normal mode for the adsorbate layer. The constant A contains all factors that do not depend on $\vec{k}$, $\vec{k}'$, or $\vec{q}$.

To obtain the inelastic current through the barrier, the matrix element in (1) is squared and summed over all initial states, all final states, and all normal modes. Due to the conservation of wave vector across the junction as indicated by the δ-function in (1), the sum over the final states can be eliminated in favor of the sum over normal modes. Therefore, the total IET intensity for a single decay energy is given by

$$I \propto \sum_{\vec{q}} F(q) \quad , \tag{5}$$

where

$$F(q) = \sum_{\vec{k}} |\langle \vec{k} | U | \vec{k} + \vec{q} \rangle |^2 \quad , \tag{6}$$

and where use has been made of the fact that the sums can be performed in either order.

The quantity $F(q)$ depends only on the magnitude of $\vec{q}$ and not on its direction (so long as the electron states are considered to be free-electron-like). For a choice of parameters typical of an Al-Al$_2$O$_3$-Pb junction [2], $F(q)$ is shown in Fig. 1. The tunneling process favors strongly transitions which have only a small momemtum change (small q) parallel to the junction interface. For larger values of q, the function approaches zero exponentially.

In order to perform the sum over $\vec{q}$, a Brillouin zone must be chosen. In the insert in Fig. 2, three Brillouin zones are shown, namely, a square, a hexagon, and a circle. The surface concentration is related to the area of the surface Brillouin zone which in turn is related to the dimension of the zones by

$$n_{sq} = q^2/\pi^2 \quad ,$$

$$n_{hex} = \sqrt{3}q^2/ 2\pi^2 \quad , \tag{7}$$

and $n_{cir} = q^2/4\pi \quad .$

The sum in (5) was performed numerically for each of the three shapes of Brillouin zone. For concentrations greater than 1% of a monolayer, the results from the three different Brillouin zones are virtually indistinguishable.

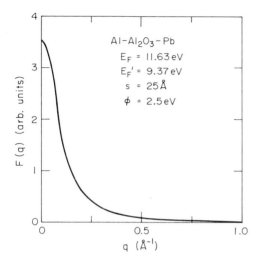

Fig.1 Function F(q) plotted against q using values typical of Al-Al$_2$O$_3$-Pb junctions. F(q) is the contribution to the IET intensity from plane wave normal modes of wave vector q.

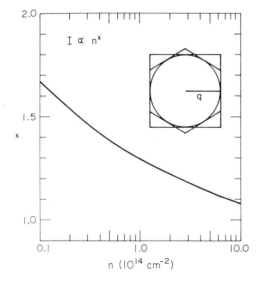

Fig.2 Exponent x as a function of concentration used in expression relating IET intensity to concentration as $I \propto n^x$. Also shown are the shapes of the three Brillouin zones used in evaluating x.

From the numerical results, the intensity as a function of concentration does not follow a strict power law. We can approximate the power law dependence by writing

$$I \propto n^x \qquad , \qquad\qquad\qquad\qquad\qquad (8)$$

where the exponent x is itself a function of n. This exponent can be obtained numerically by calculating the slope of the curve of log I versus log n. The results for all three cases are shown in Fig. 2. The results all fall within the thickness of the drafted line.

Figure 2 shows that for concentrations between 3×10^{13} cm^{-2} and 6×10^{14} cm^{-2}, the exponent varies from 1.46 to 1.12 with an average value of ~1.3. Therefore, although the intensity does not follow a power law dependence in concentration, it can be approximated closely by $I \propto n^{1.3}$ in agreement with experimental results.

Finally, it should be noted that at low concentrations, the exponent approaches the value two rather than the value unity appropriate for independent particle behavior. As the distance between adsorbate molecules becomes large, the coupling between them becomes small and the vibrations in the adsorbate layer loose their collective behavior. Consequently, our model becomes invalid in the very low coverage limit. As concentration decreases, the actual exponent will fall below the value shown in Fig. 2 and will approach the value of unity as independent vibrational behavior is achieved.

In conclusion, it has been shown that cooperative behavior between adsorbate molecules leads to an approximate power law dependence between the IET intensity and the adsorbate concentration where the exponent is greater than unity but less than 1.5. This dependence is in agreement with experimental observations.

[1] Work supported by Army Research Office (Durham) under Grant Number DAHC04-75-0170, and be the Caltech Presidents Fund PF-065.

[2] Alfred P. Sloan Foundation Fellow and Camille and Henry Dreyfus Foundation Teacher-Scholar.

References

1 J.D. Langan and P.K. Hansma: Surface Sci. 52, 211 (1975)
2 S.L. Cunningham, W.H. Weinberg, J.R. Hardy: submitted to Phys. Rev. Letters

The Golden Rule Formalism in Electron Tunneling Can It Be Justified?

T.E. Feuchtwang

Department of Physics, The Pennsylvania State University
University Park, PA 16802, USA

ABSTRACT

Recently KIRTLEY et al. [1] reported observing a marked asymmetry under bias reversal, of the inelastic tunneling spectra in metal insulator metal junctions with molecular impurities embedded in the insulator. This observation can be accounted for both within a 'golden-rule' formalism, and the conventional linear-response formulation of the transfer-Hamiltonian theory of inelastic tunneling, as developed by BENNET et al. [2]. However these two formulations differ considerably in their detailed expression for the tunneling current. This raises anew the practical question of determining the fundamental limitations, if any, on the 'golden-rule' in the electron tunneling problems in question. The golden rule formalism for inelastic tunneling is shown to be equivalent to the conventional transfer-Hamiltonian formalism as opposed to the linear response theory. Thus, the fundamental limitations on the two formulations of tunneling are equivalent. Recently CAROLI et al. [3] and FEUCHTWANG [4] have developed two alternative versions of a general and rigorous many-body formulation of tunneling which provides the necessary tool for the analysis of these limitations.

CAROLI et al. considered the contribution of vibrational excitations in the insulating barrier of a metal-insulator-metal junction, and obtained formal agreement with the 'golden-rule' [5]. A careful analysis of the same problem, within our formulation of tunneling, indicated that the tunneling current exhibits also additional terms, not accounted for by the golden-rule. These are identified with resonant tunneling phenomena which in an earlier analysis [4] were shown not to be describable by a golden-rule or transfer-Hamiltonian type formalism. These resonant contributions were apparently inadvertently missed by CAROLI et al. A discussion of consequently possible failures of the golden-rule as a phenomenological description of IETS is presented [6].

1. Introduction

There is a considerable confusion in the literature concerning the role of the golden-rule in electron tunneling theory as applied to tunneling junctions. In the following we shall examine the theoretical relation between the transfer-Hamiltonian formalism and other more recent tunneling theories to the golden-rule. We shall also consider the experimental evidence concerning the limitation on the validity of the transfer-Hamiltonian formalism. In section 2 we consider the 'derivation' of the golden-rule for tunneling phenomena in terms of the transfer-Hamiltonian.

In section 3 we consider the experimental evidence concerning the limitation on the validity of the transfer-Hamiltonian.

In section 4 we discuss possible limitations on the validity of the golden-rule formulation of tunneling which are implied by the recent tunneling
theory of FEUCHTWANG [4].

2. The Golden-Rule and the Transfer-Hamiltonian

Does the 'golden-rule' apply to the analysis of electronic tunneling phenomena in tunneling junctions? There is no simple or unqualified answer to
this question. For the answer depends on the nature of the information one
wishes to obtain from the tunneling experiment, and the nature of the justi-
fication one finds acceptable. The ordinary derivation of the golden-rule
clearly applies to tunneling phenomena which are described in terms of the
eigenstates of the complete tunneling junction. However, tunneling experiments are normally performed in order to study the physical properties of
some subsystem of the junction, such as the metal electrodes, the thin insulator, interfaces or impurities and not of the complete junction.

Therefore one is inherently concerned with the application of the golden-
rule in an approximate theory formulated in terms of the eigenstates of the
subsystems, rather than those of the entire junction. An example of this
approach is provided by CUNNINGHAM et al. [7]. To simplify the discussion,
we shall first consider the decomposition of the junction into two subsystems,
the left and right semi-infinite electrodes each of which includes the insulating barrier. The golden-rule expression for the tunneling current density flowing from left to right is

$$j = 2e \frac{2\pi}{\hbar} \sum_{\vec{k}_L, \vec{k}_R} |V_{\vec{k}_R, \vec{k}_L}(\hbar\omega)|^2 f\left(\varepsilon_L(\vec{k}_L)\right)\left[1 - f\left(\varepsilon_R(\vec{k}_R) + eV\right)\right]$$

$$\times \delta\left(\varepsilon_L(\vec{k}_L) - \varepsilon_R(\vec{k}_R) - \hbar\omega\right) \qquad (1)$$

where L(R) denote the left (right) sybsystem, whose states $|k_L>(|k_R>)$ are
labeled by the indices $\vec{k}_L (\vec{k}_R)$. $f(\varepsilon)$ is the Fermi-distribution function. A
sum over spins contributes a factor of 2. All energies are referred to the
Fermi-energy on the left and conservation of energy for the inelastic tunneling transition is assured by the Dirac-delta function. It is also assumed
that the tunneling electron can lose but cannot gain any energy in the transition. For strictly elastic tunneling one has to add the current flowing
from right to left, obtained by interchanging R and L, and thus the net
current is [8],

$$j = \frac{2e}{\hbar} \sum_{\vec{k}_L, \vec{k}_R} |V_{\vec{k}_R, \vec{k}_L}|^2 \left[f\left(\varepsilon_L(\vec{k}_L)\right) - f\left(\varepsilon_L(\vec{k}_L) + eV\right)\right] \delta\left(\varepsilon_L(\vec{k}_L) - \varepsilon_R(\vec{k}_R)\right) . \quad (2)$$

Evidently (1) asserts that the perturbation operator V induces real (as opposed to virtual) energy conserving transitions between the left and the
right subsystems.

The conventional transfer-Hamiltonian formalism was devised by BARDEEN
and others [9,10] as an explicit formal adaptation or 'derivation' of the
golden-rule for tunneling phenomena. These are viewed as an example of

quantum transitions between two 'weakly coupled' systems, each of which is characterized by its own Hamiltonian. Such a 'derivation' was necessary because the ordinary derivation of the golden-rule is invalid for quantum transitions between elements of two distinct Hilbert-spaces, such as the eigenstates $|k_L\rangle$ and $|k_R\rangle$ of the Hamiltonians $\hat{H}_L$ and $\hat{H}_R$.

BARDEEN's original analysis is restricted to the zero temperature case, and is normally interpreted to apply only to elastic tunneling [11]. In this case the 'matrix element' can be cast in the form of a 'transition current',

$$V_{\vec{k}_R \vec{k}_L} = \frac{-\hbar^2}{2m} \int_B \left[\phi_{\vec{k}_R}^*(\vec{r}) \vec{\nabla} \phi_{\vec{k}_L}(\vec{r}) - \left[\vec{\nabla} \phi_{\vec{k}_L}(\vec{r}) \right]^* \phi_{\vec{k}_R}(\vec{r}) \right] \cdot d\vec{S} \qquad . \qquad (3)$$

Here B is the boundary surface separating the left and right subsystems, and the two-dimensional integration extends over the entire surface B. In rectangular coordinates B is the surface $z = z_0$ and the integration is with respect to x and y.

DUKE [12] demonstrated that in the thick-barrier or weak-coupling limit, (1) and (3) agree exactly with the complete solution for the elastic tunneling in a noninteracting many-electron system. In this context the thick-barrier limit corresponds to the limit in which the WKB-approximation of the barrier-transmission probability consists of a single exponential. That is, the WKB-approximation of the single-particle wave functions in the barrier consists of a single decaying exponential. This result was taken by DUKE as corroboration of PRANGE's [14] original suggestion that the 'transfer-Hamiltonian' was a pseudo-operator to be used only in first order perturbation theory, and within the context of the golden-rule. That is one could conclude that subject to the above restrictions to elastic tunneling, at zero temperature between weakly coupled systems, the golden-rule applies if the matrix elements of the perturbing (pseudo) operator are calculated from BARDEEN's 'transition current'.

Formal extensions of the transfer-Hamiltonian formalism to finite temperatures and inelastic tunneling are possible. An elaborate and consistent theory was developed by BENNET, DUKE, and SILVERSTEIN [2,12]. They viewed tunneling as the linear response of the two uncoupled electrodes to the coupling or transfer-Hamiltonian. The motivation for this approach is that being linear, it inherently excludes higher order perturbation effects by the coupling-pseudo-operator. For elastic tunneling, the linear response theory reduces to,

$$j = 2e \frac{2\pi}{\hbar} \sum_{\vec{k}_L, \vec{k}_R} |V_{\vec{k}_R \vec{k}_L}|^2 \int_{-\infty}^{\infty} d\varepsilon [f(\varepsilon) - f(\varepsilon + eV)] \rho_L(\vec{k}_L, \varepsilon) \rho_R(\vec{k}_R, \varepsilon + eV), \quad (4)$$

and for a noninteracting particle the spectral density $\rho(\vec{k}, E)$ reduces to a delta function,

$$\rho_\alpha(\vec{k}, \varepsilon) = \delta \left\{ \left[\varepsilon_\alpha(\vec{k}_\alpha) + E_{F_L} - E_{F_{L\alpha}} \right] - \varepsilon \right\} \; ; \; \alpha = L, R \qquad . \qquad (5)$$

131

where, E_{F_α} is the Fermi energy of the αth electrode. Equations (3)-(5) are clearly equivalent to (2) and (3). Thus the linear response theory and the conventional transfer-Hamiltonian-golden-rule formalism are equivalent as far as elastic tunneling is concerned. Therefore, the linear response seems to provide a reasonably straightforward, systematic, and reliable extension of the transfer-Hamiltonian formalism to inelastic tunneling at finite temperatures. Unfortunately, when it is applied to inelastic tunneling processes, the linear response theory is no longer equivalent to the purely formal application of the golden-rule. This is illustrated by one of the simpler examples of inelastic tunneling: The one-phonon assisted tunneling. Here, the linear response theory gives, for the net current,

$$ j = 2e \frac{2\pi}{\hbar} \sum_{\vec{k}_L, \vec{k}_R} |V_{\vec{k}_p, \vec{k}_L}(\vec{p})|^2 \int_{-\infty}^{\infty} d\varepsilon_L d\varepsilon_R \left\{ \left[f(\varepsilon_L) - f(\varepsilon_L + eV) \right] \right. $$

$$ \times \left. \left[N(\varepsilon_L - \varepsilon_R + eV) + 1 - f(\varepsilon_R) \right] \rho_L(\vec{k}_L; \varepsilon_L) \rho_R(\vec{k}_R, \varepsilon_R) \sigma(\vec{p}, \varepsilon_L - \varepsilon_R + eV) \right\} . \quad (6) $$

Here the energies ε_α are now referred to E_{F_α}. The Bose-Einstein distribution,

$$ N(\hbar\omega) = \left[\exp \frac{\hbar\omega}{k_B T} - 1 \right]^{-1} \quad , \quad (7) $$

and $\sigma(\vec{p}, \hbar\omega)$ is the phonon spectral density. If the electrons are approximated by a noninteracting electron gas, and the phonons can be handled within the harmonic approximation then,

$$ \rho(\vec{k}, \varepsilon) \rightarrow \delta\left[\varepsilon - \varepsilon(\vec{k}) \right] \quad , \quad (8) $$

and

$$ \alpha(\vec{p}; \varepsilon) \rightarrow - \left[\delta\left[\varepsilon + \hbar\omega(\vec{p}) \right] - \delta(\varepsilon - \hbar\omega(\vec{p})) \right] \quad (9) $$

In the zero-temperature limin $N \rightarrow 0$, and the electrons cannot gain any energy but only lose energy by phonon emission. Therefore only the second term in (8) is to be retained unless $\hbar\omega(\vec{p}) = 0$. In particular, one can use (8) and (9) in (6) to determine the effect of a single impurity-phonon of energy $\hbar\omega(\vec{p})$, which is localized in the barrier. The inelastic tunneling current is, in this case given by,

$$j_{\text{non-inter.}}^{(T=0)} = 2e \frac{2\pi}{\hbar} \sum_{\vec{k}_L, \vec{k}_R} \left\{ |V_{k_R k_L}(\hbar\omega)|^2 \left[f\left(\varepsilon_L(\vec{k}_L)\right) - f\left(\varepsilon_L(\vec{k}_L) + eV\right) \right] \right.$$

$$\left. \times \left[1 - f\left(\varepsilon_R(\vec{k}_R) + eV\right) \right] \delta\left(\varepsilon_L(\vec{k}_L) - \varepsilon_R(\vec{k}_R) - \hbar\omega\right) \right\} \tag{10}$$

Here again the energies $\varepsilon_\alpha(\vec{k}_L)$ are referred to E_F. In (10) the matrix element $V_{k_R k_L}(\hbar\omega)$ is that of the electron-phonon interaction, and is not that given by BARDEEN's transition current. One may note however, that formally in the limit $\hbar\omega \to 0$ (10) reduces to (1) provided one interprets $V_{k_R k_R}(\hbar\omega=0)$ as BARDEEN's matrix element for elastic processes. This is, however, not necessarily meaningful, since as already noted the spectral density σ vanishes at $\hbar\omega = 0$, and thus the correct limit of (9) is zero. Equation (10) differs significantly from the golden-rule type expression given by (1). The two expressions predict the same peak location and the asymmetry under bias reversal of the differential conductivity, dJ/dV^2. However, they differ in their prediction concerning the somewhat smoother background on which these peaks are superposed. Furthermore, the concudtivity predicted by the linear response theory is evidently unreasonable. These conclusions are easily derived from (1) and (10) with the help of the transformation indicated below, which converts the three-dimensional sum over $\vec{k}$ to an integration with respect to energy ε and a two-dimensional integration with respect to $d^2S(\vec{k})$ ranging over the constant energy surfaces $\mathcal{S}(\varepsilon)$,

$$\sum_{\vec{k}} \; \rangle \; \int d\varepsilon \int_{\mathcal{S}(\varepsilon)} \frac{d^2 S(\vec{k})}{|\nabla_k \varepsilon|} \tag{11}$$

Substituting (11) into (1) and (10) we obtain for the tunneling conductivity in the linear-response theory,

$$\left. \frac{dj}{d(eV)} \right|_{\text{lin-resp}} = \frac{2e}{\hbar} \int_{\mathcal{S}_L(-eV+\hbar\omega)} \frac{d^2 S_L}{|\nabla_k \varepsilon_L|} \int_{\mathcal{S}_R(-eV)} \frac{d^2 S_R}{|\nabla_k \varepsilon_R|^2} |V_{\vec{k}_R \vec{k}_L}(\hbar\omega)|^2$$

$$\times \left[f(\hbar\omega - eV) - f(\hbar\omega) \right]$$

+ smooth (background) terms $\tag{12}$

In the expression for the conductivity obtained from the golden-rule the factor $[f(\hbar\omega-eV)-f(\hbar\omega)]$ is replaced by the single Fermi function $f(\hbar\omega-eV)$.

133

A similar calculation gives for the differential tunneling conductivity for both theories the following expression,

$$\frac{d^2j}{d(eV)^2} = \frac{2e}{\hbar} \int_{\mathcal{S}_R(-eV)} \frac{d^2S_R}{|\nabla_k \varepsilon|} \int_{\mathcal{S}_L(0)} \frac{d^2S_L}{|\nabla_k \varepsilon|} |V_{\vec{k}_R \vec{k}_L}(eV)|^2 \delta(eV-\hbar\omega)$$

$$+ \text{ less singular terms} \qquad . \tag{13}$$

Here all energies are referred to the left hand Fermi-energy. Equation (13) shows that the inelastic tunneling spectrum is determined primarily by transitions from the left Fermi-surface to the right Fermi surface.

The asymmetry in the differential conductivity is due to the fact that in the matrix element is calculated between two states $|k_L>$ and $|k_R>$ whose energies are respectively on the left and right Fermi surfaces, which are displaced relative to each other by the bias energy, $eV = \hbar\omega$. The asymmetry of the transmission probability (or matrix element) for inelastic tunneling can be visualized in terms of the WKB-approximation. In this approximation the transmission probability for tunneling at an energy ε across a barrier extending from x_1 to x_2 is,

$$T_{WKB}(\varepsilon) = \exp \left\{ -2 \left(\frac{2m}{\hbar^2}\right)^{\frac{1}{2}} \int_{x_1}^{x_2} [V(x)-\varepsilon]^{\frac{1}{2}} dx \right\} . \tag{14}$$

Thus, if an impurity located at R_0 can absorb an energy $\hbar\omega = eV$ then the inelastic transmission probability across a symmetric barrier of width W is given by,

$$T_{WKB,LR}$$

$$= \exp \left\{ -2 \left(\frac{2m}{\hbar^2}\right)^{\frac{1}{2}} \int_{\frac{-W}{2}}^{R_0} [V(x)-\varepsilon]^{\frac{1}{2}} dx \right\} \exp \left\{ -2 \left(\frac{2m}{\hbar^2}\right)^{\frac{1}{2}} \right.$$

$$\left. \times \int_{R_0}^{\frac{W}{2}} [V(x)-(\varepsilon-\hbar\omega)]^{\frac{1}{2}} dX \right\} . \tag{15}$$

Under bias reversal the current flows from right to left. An asymmetry in the tunneling is indicated by,

$$\ln[T_{LR}/T_{RL}]$$

$$= -2 \sqrt{\frac{2m}{\hbar^2}} \int_{-R_0}^{R_0} \left\{ [V(x)-\varepsilon]^{\frac{1}{2}} - [V(x)-(\varepsilon-\hbar\omega)]^{\frac{1}{2}} \right\} dx \neq 0 . \tag{16}$$

One may take the area underneath the curve $V(x)-\varepsilon$ as a measure of the WKB exponent. This allows the simple schematic illustration of the above argument, depicted for a square barrier in Fig. 1 below.

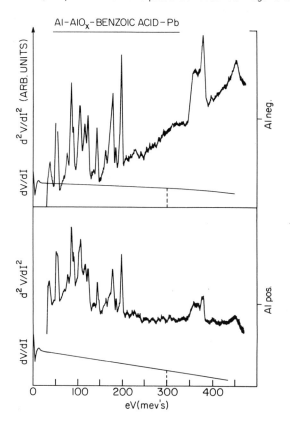

Fig.1 Typical IETS spectrum: differential tunneling resistivity and dynamical resistivity of doped tunneling junction. (After KIRTLEY et al. [1]). Note that for technical reasons KIRTLEY et al. preferred to measure the differential resistance rather than the differential conductance. These quantities are essentially proportional and exhibit the same structure: $-(1/(dV/dI)^3)(d^2V/dI^2) = d^2I/dV^2$, as seen from the figure, over the range of bias considered, dV/dI is a smooth function. Note also the pronounced asymmetry of the tunneling resistivity under bias reversal: dV/dI at 300 meV increases by 22% when the polarity of the aluminum is changed from positive to negative

To conclude, in the preceding discussion we demonstrated that the transfer-Hamiltonian formalism for elastic tunneling and its extension to inelastic processes by means of the linear-response theory, do not provide an unambiguous derivation of the golden-rule formula for inelastic tunneling.

3. The Golden-Rule and Tunneling Experiments

Recent experiments reported by KIRTLEY and HANSMA [1] and by ADLER [15] exhibit a definite asymmetry of the inelastic electron tunneling spectrum obtained by second derivative techniques. A typical example of such data is shown in Fig. 2. Although the linear response theory of inelastic tunneling can account qualitatively for these phenomena, there is reason to

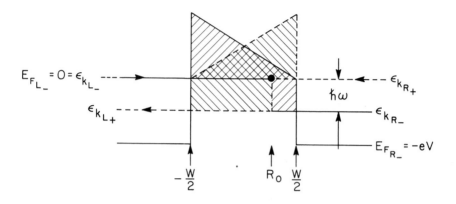

Fig.2 Schematic illustration of the asymmetry of the WKB transmission probability for inelastic tunneling under bias reversal. R_0 is the position of the molecule. The solid curve depicts the barrier with one polarity, the dashed curve depicts the barrier with reversed bias. The exponents of the WKB transmission probabilities are represented by the areas indicated by the two different shadings.

believe that it really is unsatisfactory at the relatively large values of bias at which they are observed. This conclusion follows from earlier observation that tunneling theories based on the conventional transfer-Hamiltonian formalism have a rather small range of validity. Namely, roughly the small range of external (or bias) potentials over which the elastic tunneling conductance may be approximated by its minimum value [16]. The reason for the limitation on the validity of these tunneling theories is that they represent the tunneling current as a response to an external bias potential, treated as a perturbation. That is, not only the coupling between the electrodes but also the external potential are treated within perturbation theory. However, as the bias increases, the tunneling probability increases roughly exponentially and the basic assumption of a weak coupling tends to become invalid. Thus it is evident that an adequate tunneling theory must provide for a more complete, hopefully nonperturbative treatment of the bias-dependence of the tunneling current. Recently several new tunneling theories have been proposed, all of which feature such improved treatments of the bias [3,4,17,18]. Furthermore in the thick barrier limit of elastic tunneling all reduce to the transfer-Hamiltonian formulation.

The new many-body tunneling theories can be grouped into two sets:

1) Extended-bias-function theories. These analyze the tunneling phenomena is terms of single-particle states of the complete junction, instead of the states of the several subsystems, used in the transfer-Hamiltonian formulation [17,18]. These theories could have been expressed in terms of a first order time-dependent perturbation formalism and thus forced into a golden-rule formalism. However, since they were developed in terms of a temperature-Green's function formalism these theories include corrections to the golden-rule.

2) Theories based on KELDYSH's perturbation theory [19]. These theories are developed in terms of real-time-dependent Green's functions. Thus they also go beyond first order time-dependent perturbation theory. However, in contrast to the first set of theories, this second set preserves the point of view of the transfer-Hamiltonian formalism in describing the tunneling current as due to a coupling between rigorously defined subsystems [3,4]. This feature assures that within certain limitations the tunneling characteristics can be interpreted in terms of the intrinsic properties of the subsystems rather than those of the complete junction. Again, as we shall show in more detail, these theories do not necessarily lead to a golden-rule type formulation of tunneling.

In contrast to the theories developed by CAROLI et al. [3] and by FEUCHTWANG [4] the usefulness of the extended-basis-function theories is inherently limited, due to their formulation in terms of extended basis function. This point and a more general critique of the extended-bias-function theories were considered by FEUCHTWANG [6]. Consequently the following discussion will be restricted to a review of some important results of the theories of CAROLI et al. and FEUCHTWANG.

4. Tunneling Theories Based on KELDYSH's Perturbation Theory and the Golden-Rule
4.1 Derivation of the Conventional Transfer-Hamiltonian for Inelastic Tunneling

The basic idea of the theories based on KELDYSH's perturbation theory is to treat the bias potential exactly in terms of a matrix-Dyson-equation derived by KELDYSH for the retarded and advanced Green's functions $G^{r,a}$ and the pair of correlation functions $G^{\pm}$ for systems which are not in thermal equilibrium. The statistical average of the current is expressed in terms of the correlation function G^{+}, using the relation

$$\langle j(\vec{r},t)\rangle = -2e \; \frac{\hbar}{2m} \left[\nabla_r - \nabla_{r'}\right] G^{+}(\vec{r},t;\vec{r}'t') \bigg|_{\substack{r=r' \\ t=t'^{+}}} , \qquad (17)$$

where

$$G^{+}(\vec{r},t;\vec{r}',t') = i < \psi^{\dagger}(\vec{r}',t') \; \psi(\vec{r},t)> = [G^{-}(\vec{r}',t';\vec{r},t)]^{*} ,$$

and $\psi, \psi^{\dagger}$ are field operators.

In discussing the implication of our theory concerning the validity of the golden-rule, we shall require the results summarized below,

BARDEEN's transition current operator corresponds to a classical pseudo operator which assures the continuity of the logarithmic derivative of the Green's and single-particle wave function across the metal-insulator interfaces in the junction [4a]. It acts as the coupling or transfer operator in our theory [4a].

The transfer-Hamiltonian formalism derived by APPELBAUM and BRINKMAN [3] corresponds to a thick barrier limit of the more complete theory, provided that resonant tunneling phenomena can be neglected [4d], [20].

Here it should be noted that APPELBAUM and BRINKMAN [10] have suggested, without detailed proof that their derivation applies to inelastic (i.e., phonon-assisted) tunneling as well as to elastic tunneling. This is in fact correct in the sense that BARDEEN's transition current defines the matrix element for elastic tunneling as well as for inelastic tunneling, provided the state $|k_1>$ is interpreted as the inelastically scattered component of a single-particle (Bloch) state of the left electrode which is scattered by a dynamic interaction in addition to the potential scattering by the insulating barrier. As an example, one might consider the case of a single vibrator embedded in the insulator [13]. This conclusion follows from the fact that the single-particle "wave functions" $\phi_{\vec{k}_{R,L}}(\vec{r})$ which figure in BARDEEN's

transition current (3) were shown by APPELBAUM and BRINKMAN to represent spectral amplitudes, or matrix element of the field operators between the N-particle L(R) ground state and the N-1(N+1) particle state labeled by $\vec{k}_L(\vec{k}_R)$,

$$\phi_{\vec{k}_L}(\vec{r}) = <\vec{k}_L, N-1 | \psi(\vec{r}) | 0_L, N>$$

$$\phi_{\vec{k}_R}^*(\vec{r}) = <\vec{k}_R, N+1 | \psi^\dagger(\vec{r}) | 0_R, N> \qquad . \qquad (18)$$

The only restriction on these states is that the sum of their excitation energies referred to the N-particle ground states vanishes. Thus, for elastic tunneling the hole excitation energy $\varepsilon(\vec{k}_1)$ equals the particle excitation energy $\varepsilon(\vec{k}_R)$. Inelastic tunneling clearly can occur since $\psi(\vec{r})|0_1, N>$ is not an eigenstate of the exact (N-1)-particle system. We thus reach the important conclusion that if resonant tunneling phenomena are neglected, then APPELBAUM and BRINKMAN's formulation of the transfer-Hamiltonian and consequently also the golden-rule for inelastic tunneling corresponds to a thick barrier limit of the more complete theory.

4.2 The Hamonically Vibrating Barrier

The general conclusion concerning the limitations on the validity of the golden-rule in tunneling is illustrated quite clearly by the analysis of the inelastic tunneling across a harmonically vibrating one-dimensional barrier [2b]. In this case the tunneling current can be expressed as a sum of two terms,

$$j = j_{n \cdot r} + j_{res} \qquad . \qquad (19)$$

The nonresonant contribution agrees with the golden-rule in the sense that it depends only on the product of the local densities of states $\rho(z,\varepsilon)$ evaluated at the metal-insulator interfaces at $z = L,R$:

$$j_{n \cdot r} = \frac{2e}{h} \int_{-\infty}^{\infty} \left| \frac{\hbar^2}{2m} \Gamma^r(L,R;\varepsilon,\varepsilon_1) \right|^2 \left[f(\varepsilon) - f(\varepsilon_1 + eF) \right]$$

$$\times \, \rho_L(L;\varepsilon) \rho_R(R;\varepsilon_1 + eV) \, d\varepsilon \, d\varepsilon_1 \quad . \tag{20}$$

Here the 'matrix element' $\Gamma^r(L,R;\varepsilon,\varepsilon_1)$ is given by

$$\Gamma^{r,a}(L,R;\varepsilon,\varepsilon_1) = \frac{-\hbar^2}{2m} \frac{\partial^2}{\partial z \partial z_1} G^{r,a}(z,z_1;\varepsilon,\varepsilon_1) \Bigg|_{\substack{z=L \\ z_1=R}} \quad . \tag{21}$$

If the barrier potential has a dynamic component, vibrating at the frequency ω, then the energy dependence of the matrix element is indicated below,

$$\Gamma^r(L,R;\varepsilon,\varepsilon_1) = \sum_{n \neq 0} \gamma^r(L,R;\varepsilon,\varepsilon_1)\delta(\varepsilon - \varepsilon_1 - n\hbar\omega) + \Gamma^r(L,R;\varepsilon)\delta(\varepsilon - \varepsilon_1) \quad . \tag{22}$$

Here, $\Gamma^r(L,R;\varepsilon)$ is defined by a relation analogous to (21). In particular, when the barrier is static, then $j_{n \cdot r}$ reduces to the ordinary elastic tunneling current,

$$j_{n \cdot r} = \frac{2e}{h} \int_{-\infty}^{\infty} \left| \frac{\hbar^2}{2m} \Gamma^r(L,R;\varepsilon) \right|^2 [f(\varepsilon) - f(\varepsilon + eV)] \rho_L(L,\varepsilon) \rho_R(R,\varepsilon + eV) \, d\varepsilon \quad . \tag{23}$$

The resonant contribution to the tunneling current includes both elastic and inelastic processes, for the sake of clarity we shall indicate only the resonant elastic term,

$$j_{res} = \frac{e}{h} \int \left\{ \left[f(\varepsilon) - f\left(\varepsilon - E_{F_B} + E_{F_L}\right) \right] \rho_L(L,\varepsilon) \right.$$

$$\left. \times \sum_{\alpha,\beta} \left(\frac{\hbar^2}{2m}\right)^2 \Gamma^r(L,\alpha;\varepsilon) \, \rho_{B_0}\left(\alpha,\beta;\varepsilon - E_{F_B} + E_{F_L}\right) \Gamma^a(\beta,L;\varepsilon) \right\} \, d\varepsilon$$

$$+ \text{ the same expression with L and R interchanged} \quad . \tag{24}$$

Here the indices α,β run over L and R, and

$$\rho_{B_0}(z,z';\varepsilon) = i \left[G^r_{B_0}(z,z';\varepsilon) - G^a_{B_0}(z,z';\varepsilon) \right] \quad , \tag{25}$$

is the spectral density in the barrier induced by an impurity or by the static potential representing an impurity. It is clear, that (24), (25) cannot be reduced to a golden-rule type expression such as (1) or (20). The resonant tunneling current depends on the local densities of states of the left and right electrodes in a manner which differs characteristically from the simple dependence of the nonresonant current on these quantities.

The resonant tunneling current is to be distinguished from the effect of a possible modification of the barrier transmission coefficient by the impurity potential. This effect is lumped into the nonresonant term given by (23).

Finally, it should be emphasized that the preceding discussion was strictly one-dimensional. The three-dimensional tunneling analysis is considerably more involved [4d]. In particular, the product of the local densities of states and 'matrix element' is replaced by matrix-element-weighted averages of the corresponding spectral densities over the metal-insulator interfaces.

5. The Inelastic Tunneling Current Associated with an Impurity Phonon Localized in the Barrier

The vibrating barrier potential represents a phenomenological model of inelastic tunneling. It is therefore instructive to complement the preceding discussion by an analysis of a more fundamental model. The inelastic tunneling current associated with the vibrational excitation of an impurity in the barrier can be calculated with a reasonable effort using either CAROLI et al's. [5] or FEUCHTWANG's [6] version of the theory. Both calculations were performed subject to the following simplifying assumptions.

1) The electron-impurity-phonon interaction is completely screened at the electrode-insulator interfaces. That is, the interaction is localized within the insulating barrier.

2) The electron-phonon interaction leads to a self-energy correction of the single electron dynamics. This self-energy (nonlocal potential) operator is calculated in the MIGDAL approximation [21], and includes no vertex corrections. This is roughly equivalent to the BORN-OPPENHEIMER approximation.

3) Only single-phonon processes are considered (i.e., the calculation is linearized with respect to the electron-phonon coupling constant).

4) The junction is represented by a one-dimensional model.

The tunneling current calculated subject to the assumptions above

$$\langle j \rangle = \int_{-\infty}^{\infty} j(\varepsilon) \, d\varepsilon \quad , \tag{26}$$

can be expressed as a sum of three terms

$$j(\varepsilon) = j_{el}^{(0)}(\varepsilon) + j_{inel}^{(0)}(\varepsilon) + \Delta j(\varepsilon) \quad . \tag{27}$$

(1) The elastic term,

140

$$j_{el}^{(0)}(\varepsilon) = \frac{2e}{h}\left|\frac{\hbar^2}{2m}\right|\Gamma^r(L,R;\varepsilon)\Big|^2 [f(\varepsilon)-f(\varepsilon+eV)]\rho_L(L,\varepsilon)\rho_R(R,\varepsilon+eV) \tag{28}$$

is the 'expected' elastic current, that is the nonresonant elastic current
defined by (23).

(2) The inelastic term,

$$j_{inel}^{(0)}$$

$$= \int_{-\infty}^{\infty} d\varepsilon \int_{-\infty}^{\infty} d\varepsilon_1 \left[\int_L^R dz_1 \int_L^R dz_2 |\Lambda(z_1,z_2;\varepsilon,\varepsilon_1)|^2 \sigma(z_1,z_2;\varepsilon-\varepsilon_1) \right]$$

$$\times \left([f(\varepsilon)-f(\varepsilon_1+eV)]N(\varepsilon-\varepsilon_1) + f(\varepsilon)[1-f(\varepsilon+eV)] \right) \rho_L(L;\varepsilon)\rho_R(R;\varepsilon_1+eV) \tag{29}$$

is the 'expected' inelastic current, that is, it reduces to the golden-rule
expression, (1) if:

(i) the zero temperature limit is taken, so that Bose distribution $\mathcal{N}$
tends to zero.

(ii) The electrons are 'free' i.e.,

$$\rho_\alpha(\alpha,\varepsilon) = \sum_{k_\alpha} |\psi_{k_\alpha}(\alpha)|^2 \delta\Big[\varepsilon-\varepsilon_\alpha(k_\alpha) - (E_{F_L}-E_{F\alpha})\Big] \quad ; \quad \alpha = L,R \quad . \tag{30}$$

(iii) The impurity phonon is harmonic, i.e.,

$$\sigma(z_1 z_2;\varepsilon-\varepsilon_1) = \phi(z_1)\phi(z_2)[\delta(\varepsilon-\varepsilon_1-\hbar\omega)-\delta(\varepsilon \varepsilon_1+\hbar\omega)] \quad . \tag{31}$$

The 'matrix element' $\Lambda(z_1,z_2;\varepsilon,\varepsilon_1)$ is in fact a complicated functional of
the Green's functions $G^{r,a}$, defined below,

$$|\Lambda(z_1,z_2;\varepsilon,\varepsilon_1)|^2$$

$$= \left[\int_L^R dz_3 \frac{\partial G_0^r}{\partial}(z,z_3;\varepsilon) V(z_3 z_1) \frac{\partial G_0^r}{\partial z'}(z_3,z';\varepsilon_1) \right]\Bigg|_{\substack{z=L\\z'=R}}$$

$$\times \left[\int_L^R dz_4 \frac{\partial G_0^a}{\partial z'}(z',z_4;\varepsilon_1) V(z_2,z_4)\frac{\partial G_0^a}{\partial z}(z_4,z;\varepsilon) \right]\Bigg|_{\substack{z=L\\z'=R}} \quad . \tag{32}$$

V(z,z') is the electron phonon interaction, and the subscript 'o' on the Green's functions, indicates that they are to be calculated neglecting the electronic self-energy due to the electron-phonon interaction.

(3) The correction term

The correction terms, lumped into Δj contribute both to the elastic and in-elastic single-phonon tunneling currents. We have not completed the analysis of these corrections, which are presented in an earlier publication [6]. How-ever, we have been able to identify unambiguously some terms representing resonant phenomena, in the sense discussed earlier in this section. That is, these contributions to the tunneling current depend on the local densities of states of the electrodes in the manner characteristic of resonant tun-neling. Furthermore, the spectral density in the gap is dynamically induced by the impurity. That is, it vanishes when the electron-phonon interaction is 'switched off'. This admittedly incomplete analysis provides an interest-ing example of the general limitations on the validity of the golden-rule formalism, discussed in the preceding section.

6. Conclusions

We have examined several approaches for testing the limitations on the valid-ity of the two related, golden-rule and transfer-Hamiltonian, formulations of tunneling. We have indicated that in IETS one should discuss the tunnel-ing processes in terms of well-defined subsystems, such as electrodes, bar-riers, interfaces, etc. Such a formulation is possible, within limitations, in terms of the phenomenological golden-rule. However, if a detailed theo-retical interpretation of the spectrum is required, then a more complete and admittedly more complicated theory, such as ours, probably has to be applied. The theory which we described has the potential of providing a basis for a detailed interpretation of IETS in terms of the properties of the physically interesting subsystems of the junction.

We have presented a formal, general proof for the inability of the transfer-Hamiltonian and golden-rule type formulations to describe resonant tunneling events, typically due to impurity-induced localized densities of states in the barrier. We have illustrated that for the particular case of impurity-phonon-inelastic tunneling processes, a microscopic model indeed indicated terms in the tunneling current which represent resonant effects, and thus cannot be fitted into a golden-rule type formalism. These terms apparently were inadvertently dropped by CAROLI et al. [5].

In conclusion we wish to emphasize the following two points:

1) It may be unwarranted and misleading to interpret every or even most of the IETS structures within a strictly golden-rule formalism, which as we saw cannot describe in principle all structures of this spectrum because it fails to account for resonant processes.

2) A junction does inherently exhibit excitations which are characteristic of the entire junction and not of any particular component subsystem. Such excitations clearly cannot be described by the golden-rule formulation dis-cussed in this paper.

References

1   J. Kirtley, P.K. Hansma: Phys. Rev. B13, 2110 (1976), and J. Kirtley,
    D.J. Scalapino, P.K. Hansma: Phys. Rev. B14, 3122 (1976)
2   A.J. Bennet, C.B. Duke, S.J. Silverstein: Phys. Rev. 176, 969 (1968).
    Also C.B. Duke: Tunneling in Solids, Solid State Phys. Suppl. (Academic,
    New York, 1969), Vol. 10
3   (a) C. Caroli, R. Combescot, P. Nozières, D. Saint-James: J. Phys. C4,
    916 (1971); (b) C. Caroli, R. Combescot, D. Lederer, P. Nozières, D.
    Saint-James: J. Phys. C4, 2598 (1971); (c) R. Combescot: J. Phys. C4,
    2611 (1971); (d) C. Caroli, R. Combescot, P. Nozières, D. Saint-James:
    J. Phys. C5, 21 (1972)
4   T.E.Feuchtwang: (a) Phys. Rev. B10, 4121 (1974); (b) B10, 4235 (1974);
    (c) B12, 3979 (1975); (d) B13, 517 (1967)
5   See section 4 in ref. 4d
6   A preliminary discussion of several of these questions is given by T.E.
    Feuchtwang: International Journal of Quantum Chemistry, Proceedings of
    the 1977 Sanibel Symposium. (To be published.) The discussion of the
    linear-response theory in this publication is in error and should be dis-
    regarded. The same applies to the caption of Fig. 3
7   S.L. Cunningham, W.H. Weinberg, J.R. Hardy paper CD4 in these proceedings
8   Occasionally, it is preferable to refer the right hand structure energy
    $\varepsilon_R(\vec{k}_R)$ to the right Fermi level and to express conservation of energy by
    the factor $\delta(\varepsilon_L(k_L) - \varepsilon_R(k_R)-eV)^2$
9   (a) J.R. Bardeen: Phys. Rev. Lett. 6, 57 (1961); (b) M.H. Cohen, L.M.
    Falikov, J.C. Phillips: Phys. Rev. Lett. 8, 316 (1962)
10 J.A. Appelbaum, W.F. Brinkman: Phys. Rev. 186, 464 (1969)
11 This point is discussed by Duke [12] in his excellent review of tunneling
    experiment and theory prior to 1968. However Schrieffer [13] has suggest-
    ed that the analysis applies also to inelastic tunneling
12 C.B. Duke: Tunneling in Solids, Suppl. 10, Solid State Physics, ed. by
    F. Seitz, D. Turnbull, H. Eherenreich, (Academic Press, New York, 1969)
13 J.W. Davenport, W. Ho, J. Kirtley, J.R. Schrieffer: paper CC1 of these
    proceedings
14 R.P. Prange: in Lectures on the Many-Body Problem, ed. by E.R. Caianello,
    (Academic Press, New York, 1964), Vol. 2, p. 137
15 Adler, J.G.: paper CD1 of these proceedings
16 J.M. Rowell, W.L. McMillan and W.L. Feldman: Phys. Rev. B8, 5875 (1973)
17 C.B. Duke, C.C. Kleiman, T.E. Stakelon: Phys. Rev. B6, 2389 (1972)
18 (a) W. Schattke, G.K. Birkner: Z. Physik 252, 12 (1972); (b) G.K. Birkner,
    W. Schattke: Z. Physik 256, 185 (1972)
19 Keldysh, L.V.: Zh. Eksp. Theo. Fiz 47, 1518 (1964), [Sov. Phys.-JETP 20,
    1018 (1965)]
20 Note that the derivation in ref. 4d assumes elastic tunneling. The proof
    can be repeated for inelastic tunneling, using the results of ref. 2b,
    but only if resonant tunneling (elastic or otherwise) is excluded
21 Migdal's approximation or theorem is discussed in A.L. Fetter, J. D.
    Walecka, Quantum Theory of Many Particle Systems (McGraw-Hill, New York,
    1971), p. 402

Calculations of Inelastic Tunneling Cross Sections Using Self-Consistent Multiple Scattering Techniques[1]

J.W. Davenport, W. Ho, J. Kirtley, and J.R. Schrieffer

Department of Physics, University of Pennsylvania
Philadelphia, PA 19174, USA

ABSTRACT

In the theory of inelastic electron scattering from molecules, it is known
that the distortion of the initial and final state electronic wavefunction
caused by elastic scattering plays an important role. Strong resonances
in the inelastic cross section which mirror resonances in the elastic scattering dramatically illustrate this effect. We have developed a multiple
scattering approach to include such elastic scattering effects. The scattered
wave muffin-tin potential method is used to calculate the scattering amplitude f for fixed nuclear positions. "f" then acts as an effective potential
for the inelastic scattering process. Multiphonon processes are obtained
in addition to single phonon transitions. The theory is being applied both
to tunneling and positive energy scattering processes and results will be
discussed.

---

[1] Work supported by the NSF Materials Research Division and the American Gas
Association

IV. Discussions and Comments

The Technology of IETS [1]

J.G. Adler, M.K. Konkin, and R. Magno
Department of Physics, University of Alberta
Edmonton, Canada T6G 2J1

In these remarks we shall address ourselves briefly to some of the questions which arose in previous chapters:

(1) The necessity of calibrating second harmonic signals if they are to be used for any purpose other than simply spectroscopy (such as peak intensity studies).

(2) Comments on peak asymmetry, i.e. the difference in peak magnitude for positive and negative bias direction.

(3) We will attempt to answer the question of why inelastic electron tunneling is studied primarily in Al-Pb junctions.

Several papers on instrumentation have appeared in the literature for measuring both conductance $\sigma = dI/dV$ and a signal which is related to the second derivative $d\sigma/dV$ of tunnel junctions. These are discussed in the review articles by KEIL, GRAHAM and ROENKER [1] and by COLEMAN, MORRIS and CHRISTOPHER [2]. In many measurements one is only interested in the spectra (i.e. the energy at which a particular vibrational mode occurs) and any information about the amplitude of the peaks is ignored; for such measurements the unclaibrated second harmonic signals, $V(2\omega)$, are all that one needs. If on the other hand one needs the amplitude of $d\sigma/dV$ then the measuring instrument must be calibrated. An illustration of how gross an error can be made by not calibrating of the second harmonic output is illustrated in Fig. 1. These measurements were obtained on an updated system similar to that described by ADLER and STRAUS [3] (details of the calibration are also described in this reference). The measuring system consists of a bridge and minicomputer; two lock-in amplifiers enable simultaneous recording of the fundamental and second harmonic signals from the bridge. The actual computation from uncalibrated second harmonic output $V(2\omega)$, to absolute second derivative $d\sigma/dV$, requires about two minutes (for 2500 points of each σ and $d\sigma/dV$) and even this time is mostly input-output time on magnetic tape. The use of a minicomputer with magnetic tape or a flexible disc used as a "lab book" has a number of advantages in data handling. For example, we often normalize the data by dividing the absolute second derivative, $d\sigma/dV$, by the conductance $\sigma_0 = \sigma(V_n)$ where V_n is a specific value of bias. For junctions that we measure with both electrodes normal we usually choose $V_n = 0$.

On the other hand for junctions (e.g. Al-Pb) where the lead electrode is superconducting we usually choose $V_n = 35$ mV since we do not sweep over the low bias region because the Pb energy gap and the Pb phonons are very large compared to the inelastic structure of interest. An advantage of storing the calibrated σ and $d\sigma/dV$ on magnetic tape is that one can quickly replot

the data. One can for example display the odd or even parts of the conductance, or the logarithmic derivative, d/dV (ℓn σ), the latter two of
which are shown in Fig. 2. In the case of IETS where the important structure
is an even function of energy the even part of dσ/dV enhances this structure
while tending to cancel the random noise. The logarithmic derivative tends

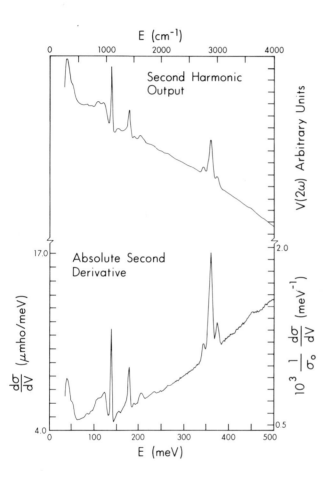

Fig.1 Raw second harmonic output of bridge, V(2ω), upper curve. Absolute
second derivative, dσ/dV, lower curve. The junction is Al-oxide-Formic
Acid-Pb Junction. The junction area is about .72 mm^2. σ_0 = σ(V$_n$ = 30 meV)

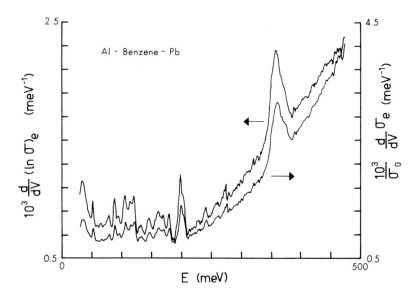

Fig.2 Comparison of second derivative and logarithmic derivatives $\sigma_0 = \sigma(30 \text{ meV})$ for an Al-C_6H_6-Pb junction having a cross section of .2mm^2

to accentuate small peaks relative to large ones, which is useful in the study of weak structures.

This brings us to the second point, namely the asymmetry of the structure. There are two types of asymmetry shown in Fig. 3. The first is just the difference in conductance in the two bias directions (Fig. 3a) and is due primarily to work function differences on either side of the barrier [4,5]. The second type of asymmetry, which was the subject of some discussion previously, is shown in Fig. 3b. We define the peak intensity:

$$F(V) = \frac{1}{\sigma_0} \int [\frac{d\sigma}{dV} - g(V)]dV \qquad (1)$$

where $g(V)$ is the smooth background $d\sigma/dV$ under the inelastic peak. We can introduce as asymmetry parameter

$$\alpha(V) = \frac{F(V_{Al})}{F(-V_{Al})} = \frac{F(-V_{Pb})}{F(V_{Pb})} \qquad (2)$$

where V_{Al} is the bias with the Al base layer positive and V_{Pb} is the bias with the Pb cover electrode positive. For example we find that for Fig. 3 $\alpha(360) = .96$. The intermediate steps in calculating α for this C-H stretching peak are shown in the curves of Fig. 4. In this particular junction there was no attempt to grow an oxide on the aluminum, but rather the aluminum was exposed for 20 min. to an ethylene plasma. With a completely homogeneous barrier one would expect no asymmetry ($\alpha = 1$). Since there is always some residual oxygen present in any vacuum system a thin oxide layer

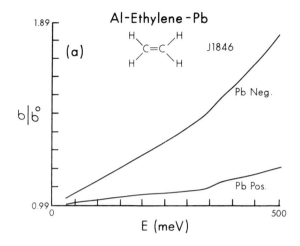

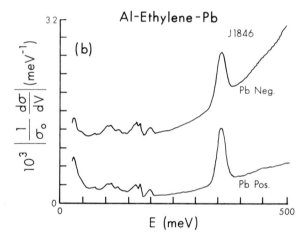

Fig.3 Showing conductance asymmetry due to work function and barrier effects (a). The lower figure (b) shows the asymmetry in peak intensities (see text).

may have been present. In any case we find that for such junctions with no oxidation step prior to depositing the organic α ≈ 1 within a few precent. If one prepares junctions in which one has Metal-Oxide-Organic-Metal then we find considerable asymmetries, as high as α ≈ .6 [6]. The reason for this is that an electron tunneling through the oxide first loses energy inelastically prior to passing through the organic, while an electron tunneling the other way can partake in all inelastic transitions available to it in the organic part of the barrier. Such two step processes in tunneling have been considered by KIRTLEY et al. [7]. In fact we find [6] that for Al-oxide plasma discharged methane-Pb junctions the tunneling becomes more asymmetric as the oxide thickness increases.

149

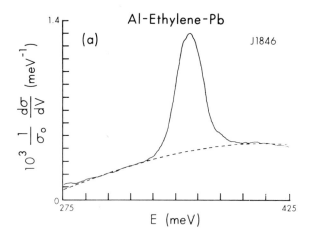

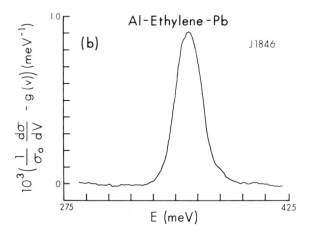

Fig.4 $(1/\sigma_0)d\sigma/dV$ for the C-H stretching mode near 360 meV with the Pb electrode biased positively shown for the junction in Fig. 3 along with the dashed curve g(V) (see text). The lower curve (b) shows integrand used in (1).

Finally we would like to make a few comments as to why Al-Pb junctions are used extensively in IETS. First let us consider the effect of the cover electrode: early work [8] indicated that the ionic radius of the cover metal plays a critical role in junction preparation. We have made a study with several cover electrodes (M_X): Ag, Pb, Sn, Au, In, and Zn. The samples examined consisted of a pair of junctions formed on a common base layer. The base was made by evaporating an aluminum film onto a clean glass slide. The barrier was produced in a DC plasma discharge of ethylene. Details of the glow discharge procedure have been reported elsewhere [9]. The two junctions sharing a common base layer and insulating barrier were completed by evaporating a Pb control electrode for the first junction and a metal M_X for the second junction. Measurements were carried out at 4.2K in liquid helium. The junctions were swept from 0 - 500 meV [M_x positive] except for those with a Pb cover layer for which the region below 35 meV was omitted to avoid saturation of the lock-in amplifiers by structure due to the Pb phonons and the superconducting energy gap. These data have been normalized at 35 meV to allow for comparison of various junctions regardless of junction resistance (the conductance at 35 meV differs from that of zero bias by less than 3% in all cases.) A typical high resolution sweep in the 275 to 425 meV range (Pb electrode positive) is illustrated in Fig. 4a, along with the dashed background curve, $g(V)$. In this discussion we limit ourselves to studying the CH stretch around 360 meV. Results for this peak are shown in Fig. 5 for the various metals. The reason for the predominance of Pb covering layers in the published literature is obvious. It is the second largest

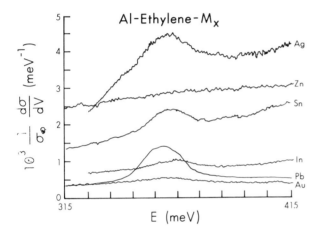

Fig.5 Typical data obtained for various covering layers

peak (next to Ag) and it has by far the best signal to noise ratio. A plot of peak intensity (for the 360 meV CH stretch) is shown as a function of ionic radius. Clearly one wants cover layers of large ionic radii for good IETS junctions, this is discussed in a forthcoming publication [10]. The reason that the signal to noise is better on Pb junctions than on Ag may be due to the granularity of the cover film [12]. Also Ag cover electrodes produce more nonlinear (steeper σ) junctions and may therefore be noisier. This aspect requires further study. Thus although the cover electrode need not be Pb it certainly gives nicer data. With regard to the base films (other than Al) a very large number of metals, among them Mg, Pb, Sn, Ta, V, Ni, Fe, Bi, Cd, Zn, Cr, Y, etc. have been used. In most of these cases oxides of the metals were used and they can often be doped with organic materials. In short, IETS does not have to be done on Al-Pb junctions, but if the problem can be studied using such junctions then it is certainly easier. An example of the use of different base layers is the subject of a forthcoming paper on the behavior of formic acid in Mg-Pb and Al-Pb junctions [11].

Finally it is interesting to note that if one examines junctions having the same covering layer, in our case Al-Ethylene-Pb, then one expects the

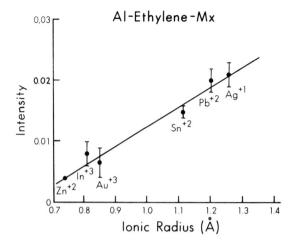

Fig.6 Intensity of the 360 meV peak as a function of ionic radius

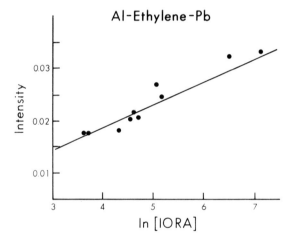

Fig.7 Intensity of the 360 meV peak plotted against ℓn10RA for a group of Al-Pb junctions

junction resistance to increase exponentially with increasing barrier thickness. Figure 7 shows a plot of the intensity of the inelastic peak at 360 meV as a function of ℓn10RA where R is the junction resistance and A its cross-sectional area. The range in areas is from $.1mm^2$ to $.7mm^2$.

The results indicate that the tunneling measurement is sampling carbon-hydrogen bonds throughout the volume of the barrier rather than just those near a metal-barrier interface.

[1] Supported in part by the National Research Council of Canada

References

1 R.G. Keil, T.P. Graham, K.P. Roenker: Applied Spectroscopy $\underline{30}$, 1 (1976)
2 R.V. Coleman, R.C. Morris, J.E. Christopher: "Experiments on Electron Tunneling in Solids" in Methods of Experimental Physics, R.V. Coleman, Ed. (Academic Press, New York, 1974) Vol. 11, p. 123
3 J.G. Adler, J. Straus: Rev. Sci. Instr. $\underline{46}$, 158 (1975)
4 W.F. Binkman, R.C. Dynes, J.M. Rowell: J. Appl. Phys. $\underline{41}$, 1915 (1970)
5 D.G. Walmsley, R.B. Floyd, W.E. Timms: Solid State Commun. $\underline{22}$, 479 (1977)
6 M.K. Konkin, R. Magno, J.G. Adler: to be published
7 J. Kirtley, D.J. Scalapino, P.K. Hansma: Phys. Rev. $\underline{B14}$, 3177 (1976)
8 A.L. Geiger, B.S. Chandrasekhar, J.G. Adler: Phys. Rev. $\underline{188}$, 1130 (1969)
9 R. Magno, J.G. Adler: Thin Solid Films $\underline{42}$, 237 (1977)
10 R. Magno, M.K. Konkin, J.G. Adler: Surface Science (in press)
11 R. Magno, J.G. Adler: to be published
12 The silver films are more matte in appearance than the lead films

Problems in the Biological Sciences

Louis Sherman

Division of Biological Sciences, University of Missouri-Columbia
Columbia, MO 65201, USA

The progress that has been made in IETS technology over the past few years
has led to the obvious quesion of whether this technique can be useful in
the study of biological molecules. However, since only a few biological
compounds have been analyzed by IETS, it is premature to attempt to answer
this question by reference to data. At this stage we can only ask more questions, and I try to determine which types of biological compounds best lend
themselves to IETS experiments. An important question which cannot be answered as yet is whether IETS may be the best technique for the study of a
particular biological problem. This must obviously wait for future developments, both theoretical and technical. From a theoretical perspective, it
would seem that IETS could be applicable to numerous biological problems.
However, the main problem at present would seem to be one of specimen preparation. These problems have been discussed by Dr. HANSMA and Dr. COLEMAN
from different points of view, and I can't add anything new to this particular discussion. I will attempt to outline a few of the areas in the study
of biological macromolecules in which IETS may be very helpful. Hopefully,
in the performance of these experiments, many of the preparative problems
can be understood and then overcome. These would not necessarily be the ultimate aims of IETS, but this approach may be helpful in learning what these
goals are and how to get there.

To date, most of the work on biological compounds has been performed on
small monomers such as amino acids, nucleotides, or lipids. In dealing with
a new technique it is obviously necessary to start at the lowest level of
organization. But, since there are numerous biochemical and biophysical
techniques that can be used on these molecules, it is unlikely that IETS will
yield unique data at this level. However, the analysis of biological macromolecules such as proteins and DNA would seem to be an area in which IETS can
make significant contributions. In particular, the three-dimensional conformation of these molecules might be an appropriate starting place for IETS.
In this way, we can very sensitively determine if such molecules can be prepared for IETS in a biologically active state.

A typical example concerns DNA which is the genetic material for all life
on earth. Dr. COLEMAN has already reported a spectra on a commercial preparation of DNA and we know that it is possible to study such a large molecule. But native DNA may be found in a variety of conformations. Some DNA
is linear, with two free ends, while DNA from certain viruses is circular.
At times, the circular DNA winds around itself, much like a twisted rubber
band, into what is called superhelical coiled DNA. Such conformations have
now been well-studied and may occur either as single or double-stranded (the
typical Watson-Crick model) forms. These molecules are usually found relatively free of protein. The DNA in the nucleus of higher organisms, on the

other hand, is tightly complexed with protein and the precise conformation of the DNA is not known with certainty. Since the structure of the DNA-protein complex is intimately related to many regulatory events this area has generated a great deal of experimental work, though without much success. Determining the structure of DNA in such a complex may be an ultimate goal of IETS research in biology.

However, for now, it would be important to see if IETS can differentiate between the different DNA conformations. An interesting test can be made using DNA isolated from bacteriophage λ, a virus which infects the bacterium, E. coli. This DNA has 10 complementary base pairs on opposite ends of the two strands. By using very simple biochemical techniques, this molecule can be converted from a linear to a circular to a superhelical circular DNA. In this way, the three conformations can be compared to each other without varying other parameters such as size. Finally, this DNA can be denatured into single strands to determine if there are spectral differences between single and double-stranded DNA. At the same time, such experiments can help answer a series of other technical questions and go a long way to determining if IETS will be useful in the study of DNA.

In discussing protein conformation, I will also refer to a molecule studied by Dr. COLEMAN - hemoglobin. This molecule, which is involved in the transport of oxygen, is actually composed of 4 individual polypeptide chains. Each polypeptide chain is similar to another oxygen-transporter, myoglobin, which exists as a single polypeptide. A comparison of the spectra of these 2 proteins would be very informative because it would decide if IETS is capable of differentiating higher orders of protein structure. Of even greater importance, it is known from X-ray diffraction studies that hemoglobin changes conformation upon binding O_2. Therefore, it would be of interest to obtain spectra from deoxyhemoglobin and oxyhemoglobin to see if such conformational changes can be detected.

As I discuss such sophisticated and valuable experiments, there is still one important question that is always in mind: Does IETS give us a spectrum of the native, active molecule or does the preparative technique denature the protein? Proteins can be treated by a variety of chemicals that convert them into an inactive form. In the extreme, the polypeptide chain resembles the random coil structure of organic polymers. Therefore, a critical experiment must be performed. A protein like myoglobin should be denatured chemically and its IETS spectrum compared with that of the "native" protein. Identical spectra would indicate either that the preparation of the protein for IETS denatures the protein into a random coil form or that the technique is unsuitable for high resolution protein studies. On the other hand, distinctly different spectra would indicate that the technique is potentially useful in the study of biological macromolecules.

These comments were just a few brief thoughts on how we may begin to determine the importance of IETS to biology. With so many unanswered questions, I could continue in this way for some time. But, I think that it would be more valuable to begin a general discussion at this time. Therefore, I would like to ask Dr. COLEMAN, who has the most experience, a few questions.

Sherman: Dr. COLEMAN, what were the concentrations that you used in your DNA studies?
Coleman: Well I don't have the exact information at the moment, but it was on the order of 1 mg per ml.

Sherman: One of the comments that I thought of when I was listening yesterday, was that the D_2O-H_2O technique, could be used to detect one molecule per 1500 square angstroms. Now if that's true it seems to me that one could work with biological macromolecules in the range of one microgram per ml.
Coleman: Yes, but there is a major problem there. Again, that is with benzoic acid which interacts with its surface and gives a very strong coupling tunneling junction. I think that everything you said about these biological molecules is potentially doable and obviously an objective of this system. The problem that I see, of course, is obtaining a sufficient mode resolution intensity to be sure that when you do the experiment you can follow those bands you are interested in. So, I think that in order to do all of the things you would like to do, we really have to find a better way to get an acceptable spectrum on these large molecules. Now, I think ideas like your suggestion with the DNA ring, and so on, to try and compare two cases ia a very good idea. We have done myoglobin and it is no different from hemoglobin. It gives you the same problems and seems to be dominated by the CH mode. We have also tried some of these enzymes with the active site. You have the same problem. The spectrum seems to be washed out and dominated by CH and it's very difficult to attempt to identify bonds that would be associated with the active site. Unless you can figure out some way to enhance those intensities.
Sherman: Do you think that using different insulator materials could be of value as far as getting the molecules to stick?
Coleman: Well, there is no problem getting the molecule to stick as soon as you go up to high molecular waves. You can obtain infinite resistance very easily. Unfortunately, that does not follow that nice curve of intensity versus resistance. In fact, that's the primary problem in the biological molecules. I don't know what the optimum resistance is but increasing the resistance does not often improve your resolution or your intensities. It often wrecks your background. I think there is a very big technical problem just trying to optimize the resolution of the tunneling.
Sherman: Had you tried lowering the concentrations of the DNA or the protein to see how the intensity fell off?
Coleman: Well, we just started doing that. I guess we spent a lot of time on the nucleotides to get oriented on how the smaller molecules were working. And it's just recently we have made an effort to explore the protein.
Sherman: Thank you.

Relation of IETS to Other Surface Studies

W. Plummer

Department of Physics, University of Pennsylvania
Philadelphia, PA 19174, USA

Today all of us are in a situation where we must find "relevance" in our research work. For surface science, catalysis has furnished an easily identi-
fiable subject which is relevant and may be impacted by basic surface science.
It has been claimed by several speakers at this conference that IETS is applicable to catalytic problems. I will not comment upon these claims, since
I do not have sufficient experience or knowledge in this area. What I will
address is the potential impact of IETS upon surface physics and chemistry.

Most scientists working in the general area of surface science would agree
that the most pressing question is where and how are foreign atoms or molecules bound to a surface? We need to know the geometry as well as the electronic structure. Given this problem, IETS would seem to be a very attractive
experimental technique to determine bonding configurations. If you can see
the vibrational modes of the chemisorbed molecular complex and can determine
the geometric orientation from the strength of the various modes you should
be able to determine the bonding configuration.

When Tom Wolfram asked me to discuss the relation of IETS to surface
science I decided that there was one crucial test of the impact of IETS on
surface science. Would I as an experimental surface scientist go home from
this meeting and set up an IETS facility; or can IETS answer the questions I
would like to know about adsorbed atoms or molecules? I have listened to the
talks given here and read the articles written by HANSMA and WEINBERG and the
answer to my own question is NO! I will try to explain why I feel this way.
Clearly IETS has tremendous promise but several major breakthroughs need to
happen before it will be a useful tool for me. My major concern is the total
lack of characterization of the interface. The surface science community has
struggled for years to make surface science a quantitative science and it
looks like we may have made it. Using IETS in its present stage of characterization would, in my view, be a step backwards. So the first major improve-
ment must be a microscopic characterization of the junction region. The
second concern of mine is the effect of the top electrode on what is in the
junction.

Let me spend a little time talking about a field which could have had an
analogous development to what may happen in IETS. The field is field ion
microscopy (FIM). When anyone first sees one of Mullers pictures you immediately think that you could solve nearly every surface problem with FIM.
You know exactly where every atom is and the small tip has every crystal orientation exposed. Many of us, including some very talented people, thought
we could solve the surface problem with FIM. In my view, only EHRLICH, with
his diffusion experiments, has broken out of field ion microscopy phenomena
and learned something fundamental about surfaces. The rest of us suffered

157

from a disease I call navelitis. You keep learning more and more about FIM, every year you got in deeper. At some stage you had to decide whether you were a field ion microscopist or a surface scientist. The same phenomena seems to be occurring in IETS. One may have a technique in search of a problem.

An example of the other extreme is photoelectron spectroscopy. When you attend a meeting on photoelectron spectroscopy, the attendees are from atomic physics, molecular chemistry, solid state physics, etc. In general, these are scientists with a problem using this technique to solve their problem. When you start to see this in IETS you will know that the technique is successful.

I can see two reasons for using IETS for surface studies. One is to use a vacuum as the insulator. Dr. Thompson of IBM will describe his difficult experiment of tunneling from a well-characterized metal through a vacuum into another well characterized metal. With this procedure I can characterize both of the interfaces with other surface techniques. The second alternative I can see is to study large transition metal molecules which have been placed inside of the barrier. Many experiments now indicate that the surface bond (at least in chemisorption) is very localized, so these molecules may be an excellent model of a surface. But in this case you must improve upon what can be learned from infrared spectroscopy.

Finally, I would like to give you a couple of examples to illustrate why I will use a different technique than IETS to study the vibration energy levels of an adsorbed molecule or atom. The experimental technique is high resolution inelastic electron scattering from the surface. I would like to illustrate the possibilities by using data obtained from gas phase molecules. The first example is from the work of SCHULTZ on C_6H_6. What SCHULTZ showed was that at different incident kinetic energies you couple to different symmetry states of the molecule. At an incident energy of 1.8 eV you couple to one symmetry set of modes and at 4.8 eV you see another symmetry set. So in first order I can overcome HANSMA's criticism of not having good enough resolution because I can tune in and out modes. The second advantage was also pointed out by SCHULTZ. At specific incident energies you form negative ion resonances. In these resonant states you can observe the first seven vibrational modes of CO or N_2. You can see the shape of the potential well for the first 1.5 eV. The independent control of the surface condition, the incident electron beam energy and direction, as well as the collection direction, seem to me to make this experiment much more useful than IETS.

V. Molecular Adsorption on Non-Metallic Surfaces

Photoemission Studies of Molecular Adsorption on Oxide Surfaces[1]

Victor E. Henrich, G. Dresselhaus[2], and H.J. Zeiger

Lincoln Laboratory, Massachusetts Institute of Technology
Lexington, MA 02173, USA

ABSTRACT

We have used ultraviolet photoemission spectroscopy (UPS), LEED and Auger spectroscopy to study the interaction of adsorbed molecules with transition-metal-oxide surfaces. Comparison of the UPS spectra of free molecules with the UPS difference spectra for adsorbed molecules, makes it possible to observe changes in the molecular electronic structure induced by adsorption. We have studied the adsorption of O_2 and H_2O on both nearly perfect and ion-beam-damaged surfaces of TiO_2 and $SrTiO_3$. The interaction of O_2 with all surfaces is similar for low exposures (≤ 100 L), with the difference spectra indicating strong, probably dissociative chemisorption. For O_2 exposures ≥ 100 L, a second adsorbed phase occurs on some surfaces. For low H_2O exposures, the difference spectra give evidence for dissociative chemisorption in some cases. For high H_2O exposures, the spectra indicate only a slight distortion of the free molecule. On TiO_2, the a_1 molecular orbital of H_2O is shifted 0.5 - 1eV toward tighter binding, suggesting that H_2O is bound to the surface via is O lone-pair orbital. On $SrTiO_3$, both the a_1 and the higher-lying b_1 molecular orbitals are shifted, suggesting a more complicated bonding. The magnitude of the extramolecular relaxation-polarization shift of H_2O on adsorption depends primarily on the presence or absence of defect electronic surface states in the bulk bandgap of the oxide.

1. Introduction

Ultraviolet photoemission spectroscopy (UPS) is a powerful tool for studying the electronic structure of surfaces and adsorbed molecules [1]. Although UPS probes the gross electronic structure of adsorbed molecules rather than the vibrational structure that is examined by inelastic electron tunneling spectroscopy (IETS), the two techniques are used to address the same questions: how do molecules bond to surfaces; what electronic changes take place upon adsorption; what is the orientation of adsorbed molecules? We will describe here some of our UPS studies on the adsorption of O_2 and H_2O on $SrTiO_3$ and TiO_2 surfaces, emphasizing the types of information that can be obtained by this method.

In UPS, photons of energy $h\nu$ are incident on a surface in vacuum. In the most probable absorption process, all the energy of a single photon is transferred to a single electron, exciting the electron to a final state at an energy $h\nu$ above its initial state. Some of the electrons whose final-state energies lie above the vacuum level are emitted into vacuum without appreciable energy loss, giving rise to a distribution of photoemitted electrons as a function of kinetic energy which, except for final-state and matrix element effects, mirrors the initial density of states. For the photon energy used here (21.2 eV), photoelectrons are emitted only from the first 10-20 Å of the sample, making UPS very sensitive to adsorbed molecules.

The electronic properties of atomically clean surfaces of $SrTiO_3$ and TiO_2 have been reported in detail previously [2]. Both materials are wide-gap semi-conductors ($E_g \approx 3.1$ eV). In the bulk, all of the Ti ions are in a $Ti^{4+}(3d^0)$ configuration; vacuum-fractured surfaces also have this configuration. When defects are produced on either material by ion bombardment (500 eV Ar ions were used in our experiments), the surfaces become disordered and lose O (and also Sr from $SrTiO_3$), leaving most of the surface Ti ions in a $Ti^{3+}(3d^1)$ configuration. These Ti^{3+} ions form a band of surface states at energies within the bulk bandgap.

2. Adsorption of O_2 on $SrTiO_3$

Figure 1 shows UPS spectra for an ion-bombarded $SrTiO_3$(100) surface before exposure to O_2 and after exposures ranging from 0.5 to 10^8 L (1 L $\equiv 10^{-6}$ Torr-sec). Each spectrum gives the number of photoemitted electrons, $n(E)$,

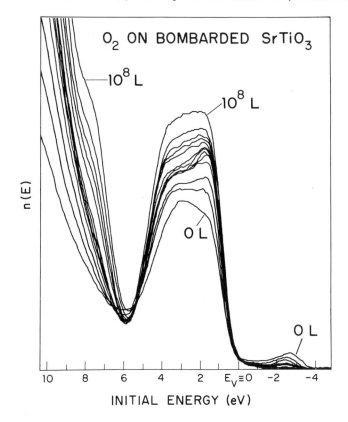

Fig.1 UPS spectra (for 21.2 eV photons) of Ar-ion-bombarded $SrTiO_3$(100) after successive exposures to O_2 (1 L $\equiv 10^{-6}$ Torr-sec)

as a function of their initial-state energy in the solid or in the adsorbed molecule. The upper edge of the valence band, E_v, is taken as the zero of initial energy. The spectrum for the atomically clean surface (0 L) shows strong emission from the O-derived valence band (0 to 6 eV) and weak emission from the Ti d-electron surface states within the bulk bandgap (-1 to -4 eV). A background of inelastically scattered electrons rises to the left (i.e., toward lower kinetic energy). The family of curves shows that significant changes occur in the UPS spectra with O_2 exposure. Emission from the d-electron surface states decreases rapidly, disappearing after an exposure of about 10 L, while emission from the region of the valence band increases. For exposures greater than 100 L a peak appears at 7-8 eV, below the valence band.

Each UPS spectrum contains information about the total electronic structure of the combined adsorbate/substrate system. The changes resulting from adsorption can be shown more clearly by plotting the difference between each spectrum and that for the clean surface; such plots are referred to as photoemission difference spectra [1]. If an adsorbed molecule interacts weakly with the substrate, the electronic distortions that occur upon adsorption can be determined by comparing the difference spectra with the UPS spectrum for the free (gaseous) adsorbate; this is the case for H_2O adsorption on $SrTiO_3$ and TiO_2, as discussed below. If there is a strong adsorbate-substrate interaction, however, the comparison procedure is in general not applicable, and the difference spectra are more difficult to interpret; this is the case for O_2 adsorption on $SrTiO_3$ and TiO_2.

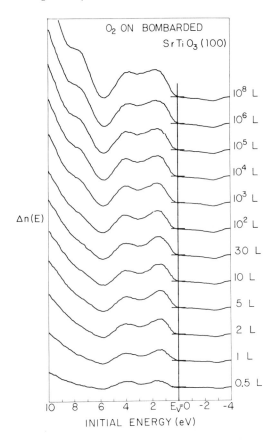

Fig.2 UPS difference spectra for Ar-ion-bombarded $SrTiO_3$ (100) after successive exposures to O_2

The difference spectra for the family of curves in Fig. 1 are plotted in Fig. 2. These spectra show that two distinct adsorbed phases are formed when ion-bombarded $SrTiO_3$ is exposed to O_2. The first phase (I), which is obtained for exposures up to about 100 L, has a sticking coefficient of 0.2-1 and gives a difference spectrum with two peaks in the region of the valence band (at about 1 eV and 4 eV). The depopulation of the d-electron surface states results in a small negative dip above E_v. The difference spectra for phase I resemble the portion of the UPS spectrum for the clean surface that is due to emission from the valence band. Since the valence band arises predominately from O^{2-} ions, we believe that O_2 is dissociatively adsorbed on ion-bombarded $SrTiO_3$ at low exposures, yielding adsorbed O^{2-} ions. A second adsorbed phase (II), with a sticking coefficient of about 10^{-3}, occurs for exposures greater than 100 L. With increasing exposure there are small changes in the relative heights of the two peaks in the region of the valence band, and a third peak appears below the bottom of the valence band. If it is assumed that phase II is adsorbed in additon to phase I, then spectra for phase II alone can be obtained by subtracting the UPS spectrum for 30 L exposure (where phase I is complete) from the spectra for 10^4 to 10^8 L. The two phases are shown in Fig. 3, where Fig. 3(a) is the difference spectrum

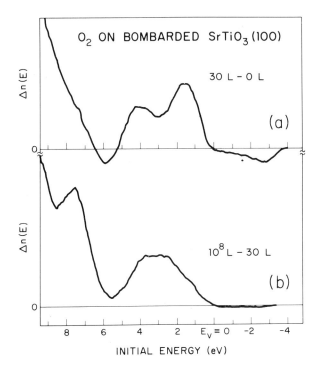

Fig.3 UPS difference spectra for Ar-ion-bombarded $SrTiO_3$(100): (a) 30 L - 0 L (phase I); (b) 10^8 L -30 L (phase II)

for phase I obtained by subtracting the spectrum for the clean surface from that for 30 L, and Fig. 3(b) shows the spectrum for phase II obtained by subtracting the 30 L spectrum from the 10^8 L spectrum. We have not yet determined the nature of phase II, although it is probably some charge state of 0 or O_2.

The difference spectra obtained when O_2 is adsorbed on vacuum-fractured $SrTiO_3(100)$, which has no d-electron surface states, are shown in Fig. 4.

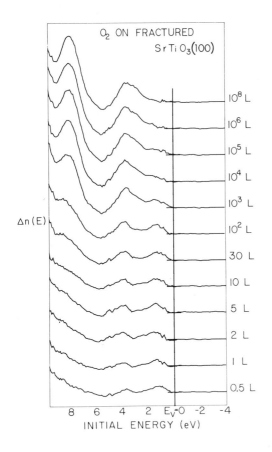

Fig.4 UPS difference spectra for vacuum-fractured $SrTiO_3$ (100) after successive exposures to O_2

Two adsorbed phases are again seen. The spectra for phase I here are essentially the same as those on ion-bombarded $SrTiO_3$, and we again attribute this phase to the dissociation of molecular O_2 into O^{2-} ions. The spectra for phase II here exhibit the same type of peak at 7-8 eV that was seen in Fig. 2, but the photoemission from the region of the valence band changes in a different way for exposures greater than 100 L. The amplitude of the peak at about 1 eV <u>decreases</u> with increasing exposure in Fig. 4, disappearing completely by $\overline{10^8}$ L, while the peak at 4 eV grows in amplitude.

The difference spectra in Fig. 4 for large O_2 exposures on vacuum-fractured $SrTiO_3$ look very much like those for phase II alone (after subtracting phase I) on ion-bombarded $SrTiO_3$, shown in Fig. 3(b). This similarity, coupled with the reduction in intensity of the peak at 1 eV on vacuum-fractured SrTiO₃ at high O₂ exposures, suggest that phase II is the same on both surfaces, but that it is adsorbed <u>in addition to</u> phase I on ion-bombarded SrTiO₃, and <u>instead of</u> phase I on vacuum-fractured surfaces.

3. Adsorption of H_2O on TiO_2 and $SrTiO_3$

The UPS spectra for adsorption of H_2O on TiO_2 and $SrTiO_3$ are somewhat easier to interpret than those for O_2 adsorption, and more information concerning the adsorbed phases can be obtained form the difference spectra. Figure 5(a)

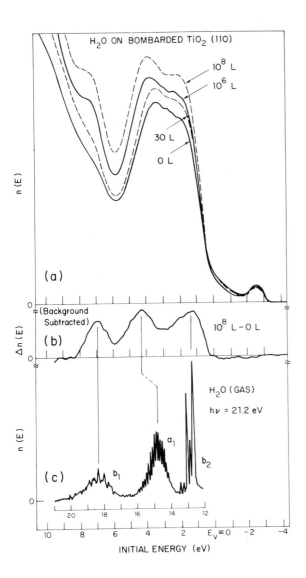

Fig.5 (a) UPS spectra for Ar-ion-bombarded TiO_2(110) after various exposures to H_2O (b) UPS difference spectrum for 10^8 L exposure. (c) UPS spectrum of free H_2O molecule (Ref. 3)

165

shows UPS spectra for ion-bombarded TiO_2(110) before exposure to H_2O and after three different H_2O exposures. Atomically clean, bombarded TiO_2(0 L) exhibits the same type of d-electron surface-state emission as does bombarded $SrTiO_3$, and the valence band emission is similar in width and general shape to that from $SrTiO_3$. Figure 5(b) shows the difference spectrum, with a smooth background subtracted, for bombarded TiO_2 after 10^8 L exposure to H_2O. Unlike O_2 exposure, H_2O exposure does not depopulate the surface states. The difference spectrum, with three peaks near 1,4 and 7 eV, is quite similar to the UPS spectrum of the free H_2O molecule [3], which is shown in Fig. 5(c). The similarity leads us to conclude that for large exposures H_2O does not dissociate on TiO_2. It is therefore possible to learn more about the nature of the adsorbate-substrate interaction by comparing the spectra from the free and the adsorbed molecules.

In making such a comparison, the two spectra are first shifted relatively in energy to align the peaks that arise from molecular orbitals that should be least affected by bonding [1]. The shift required for alignment, which is referred to as the extramolecular polarization-relaxation shift, is due to the effects that influence all the molecular orbitals equally [1]. The relative displacements of peaks in the two spectra that remain after making the polarization-relaxation shift can be attributed to distortions of the molecule upon adsorption. In Figs. 5(b) and 5(c), the b_1 and b_2 peaks have both been aligned by a polarization-relaxation shift of 3.6 eV, and the a_1 peak after adsorption is seen to be shifted 1 eV toward tighter binding. This suggests that the H_2O molecule bonds to the substrate via its a_1 orbital, which consists of the O lone-pair whose charge-density lobes bisect the H-O-H angle in the plane of the molecule [4]. A shift in this orbital suggest that the molecule bonds to the surface with its O atom down and the two H atoms oriented away from the surface. The magnitude of this shift is a measure of the strength of the adsorbate-substrate interaction.

The adsorption of H_2O on ion-bombarded TiO_2 is not as easy to interpret for low exposures as for high exposures. Figure 6 shows the compete set of difference spectra that we have obtained for this adsorbate/substrate system. For exposures below 10 L, a weak, two-peaked spectrum is seen. We believe that this spectrum is due to absorbed OH, indicating that water is dissociatively adsorbed at low exposures. From infrared spectroscopy data [5] OH is known to be a stable species on TiO_2 surfaces, and recent UPS measurements of free OH radicals [6] give a two-peaked spectrum whose separation is very close to that seen here. It is not possible from these spectra to determine whether or not the OH radicals remain when molecular H_2O is adsorbed at higher exposures.

When H_2O is adsorbed onto annealed, nearly perfect TiO_2(110) surfaces, which have no d-electron surface states, UPS difference spectra similar to those in Figs. 5 and 6 are obtained. The same adsorbed phase is present at low exposures, and the a_1 orbital is again shifted relative to the free molecule at high exposures. The adsorbate-substrate interaction seems to be weaker on annealed TiO_2, however, since the a_1 orbital is shifted by only 0.7 eV relative to the free molecule.

Our UPS results for the adsorption of H_2O on ion-bombarded $SrTiO_3$(100) are shown in Fig. 7. The difference spectrum for 10^8 L exposure, Fig. 7(b), exhibits three peaks below E_v as well as a negative dip above E_v corresponding to depopulation of the d-electron surface states. (This depopulation occurs only for exposures greater than 10^5 L, however, in contrast to the case of

166

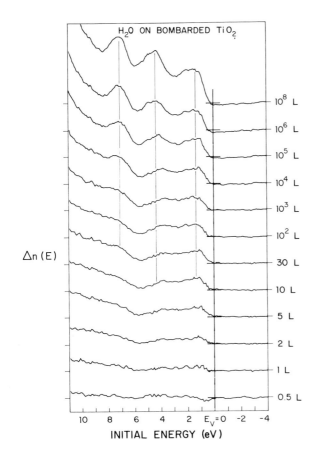

H₂O ON BOMBARDED TiO₂

$\Delta n (E)$

10⁸ L
10⁶ L
10⁵ L
10⁴ L
10³ L
10² L
30 L
10 L
5 L
2 L
1 L
0.5 L

10   8   6   4   2 $E_V = 0$ -2   -4

INITIAL ENERGY (eV)

Fig.6 UPS difference spectra for Ar-ion-bombarded TiO_2(110) after successive exposures to H_2O

O_2 adsorption.) The separation between the b_1 and b_2 peaks in the difference spectrum is less than in the free molecule spectrum, Fig. 7(c). We have a-ligned the b_1 peaks in the two spectra because this orbital should be least perturbed by bonding. With this alignment, the a_1 and b_2 orbitals in the adsorbed spectrum are both seen to be shifted 0.5 eV toward tighter binding, suggesting that both the in-plane O lone-pair orbital (a_1) and the O lone-pair orbital normal to the plane of the molecule (b_2) [4] are involved in the bonding. A simple interpretation of the orientation of the adsorbed molecule is not apparent in this case.

In conclusion, we have presented UPS results showing that this technique is a powerful tool for the study of adsorbed molecules that gives information complementary to that obtained from IETS.

167

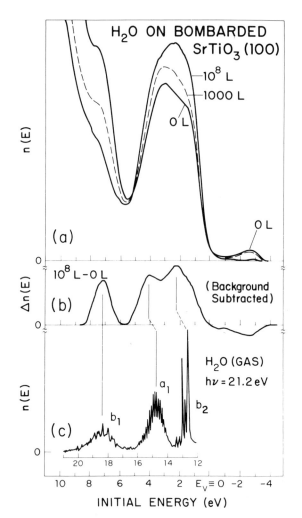

Fig.7 (a) UPS spectra for Ar-ion-bombarded $SrTiO_3$ (100) after various exposures to H_2O. (b) UPS difference spectrum for 10^8 L exposure. (c) UPS spectrum of free H_2O molecule

Acknowledgments

The authors thank E.I. Solomon for helpful discussions and B. Feldman for technical assistance.
The views and conclusions contained in this document are those of the contractor and should not be interpreted as necessarily representing the official policies, either expressed or implied, of the United States Government.

[1] This work was sponsored by the Department of the Air Force
[2] Present address: Francis Bitter National Magnet Laboratory, M.I.T.

References

1   J.E. Demuth, D.E. Eastman: Phys. Rev. Lett. $\underline{32}$, 1123 (1974)
2   V.E. Henrich, G. Dresselhaus, H.J. Zeiger: Phys. Rev. Lett. $\underline{36}$, 1335 (1976)
3   D.W. Turner, C. Baker, A.D. Baker, C.R. Brundle: <u>Molecular Photoelectron Spectroscopy</u> (Wiley-Interscience, New York, 1970), p. 113
4   C.J. Ballhausen, H.B. Gray: <u>Molecular Orbital Theory</u> (Benjamin, New York, 1965), p. 72
5   See, for example, F. Gonzalez, G. Munuera: Revue de Chimie Minerale $\underline{7}$, 1021 (1971), and references therein
6   S. Katsumata, D.R. Lloyd: Chem. Phys. Lett. $\underline{45}$, 519 (1977)

Application of Neutron Scattering to the Study of the Structure and Dynamics of Short-Chain Hydrocarbons Adsorbed on Small Particles [1]

H. Taub
Department of Physics, University of Missouri-Columbia
Columbia, MO 65201, USA

ABSTRACT

The application of inelastic and elastic neutron scattering to the study of the dynamics and structure of adsorbed films is briefly reviewed. The experimental techniques are illustrated for the case of a butane film (CH_3 $(CH_2)_2CH_3$) adsorbed on small graphite particles. The inelastic spectrum of the butane monolayer contains new surface vibratory modes as well as the intramolecular vibrations observed in the bulk solid. A simple model fitting the principal features of the monolayer spectrum is used to predict the orientation of the adsorbed molecule and the location and strength of the bonds to the substrate. The uniqueness of this model is discussed. The use of elastic neutron diffraction to investigate the long-range order within the film and its potential for determining the orientation of the adsorbed molecules is also described. At several points in the discussion the applications of neutron scattering and IETS to molecular spectroscopy are compared and the paper concludes with suggestions for how the two techniques may interact in the future.

1. Introduction

This paper reviews some work which is in progress at the University of Missouri using neutron scattering to study the structure and dynamics of short-chain hydrocarbon molecules adsorbed on small particles. These experiments began about a year ago with an investigation of the structure and dynamics of butane adsorbed on a graphite powder, and a detailed account of this work has already appeared [1,2]. Our purpose here is to present a broader view of neutron scattering as a surface probe--one which can be followed by someone unfamiliar with the technique. We shall attempt to make the discussion relevant to the subject of this conference by comparing neutron molecular spectroscopy with IETS so that the relative strengths and weaknesses of each technique can be appreciated.

The outline of the paper is as follows: Sec. 1.1 briefly describes the types of neutron scattering experiments which can be done on adsorbed films while in Sec. 1.2 some of the key experiments which lead to the development of the technique are noted. Section 2 represents the principal part of the discussion in which the neutron technique is illustrated for the particular example of butane adsorbed on graphite. Experimental considerations such as the choice of butane as the adsorbate and graphite as the substrate will be covered in Sec. 2.1. Next, the inelastic spectra of the butane films are presented in Sec. 2.2. These spectra contain new surface vibratory modes associated with the loss of translational and rotational degrees of freedom

170

of the adsorbed molecule as well as the intramolecular excitations which are
observed in bulk butane. After comparing the bulk and film spectra, we discuss in Sec. 2.3 a simple model which we have used to fit the observed
monolayer spectrum. The aim of the modeling is to determine the eigenvectors
of the surface vibratory modes and to investigate the extent to which the
molecular orientation of the adsorbate and the location and strength of its
bonds to the substrate can be determined. Section 2.4 concludes the discussion of butane on graphite with some results of elastic neutron diffraction used to study both the molecular orientation and the long-range order
in the film. Finally, Sec. 3 summarizes our present work on other hydrocarbon films and discusses how neutron molecular spectroscopy may interact
with IETS in the future.

1.1 Neutron Scattering as a Surface Probe

As uncharged particles, neutrons interact only very weakly with matter and
therefore do not scatter preferentially from surfaces. Essentially, brute
force is used to solve this "transparency" problem. One adsorbs a strongly
scattering film onto a weakly scattering substrate of large surface area.
In practice, the surface area requirement limits one to powdered substrates.
Those which have already been used include charcoals, graphites, zeolites,
and metal oxides.

    As far as what can be learned about an adsorbed film using neutrons, the
experiments fall into two categories corresponding to elastic and inelastic
scattering. Elastic scattering probes the structure of the overlayer in
much the same way as low-energy electron diffraction (LEED). One can determine the unit cell dimensions of the film and observe structural transitions
between two-dimensional phases. An advantage of neutron scattering over LEED
is that the penetrability of the neutron allows one to work with films at
high vapor pressure so that ultra-high vacuum is not required.

    In the case of inelastic scattering, neutrons can be used to study the
dynamical properties of films in much the same way as in their application
to excitations in bulk matter. The energy range accessible to thermal neutron scattering is generally less than 0.1 V so that the range is more restricted than in IETS but complements it on the low-energy side. While neutron scattering has neither the energy range nor sensitivity of IETS, it does
have several compensating advantages. Because both the momentum and energy
transfer of the neutron can be varied in an inelastic scan, one can observe
the Q-dependence as well as the density of states of the excitations under
investigation. The neutron-phonon cross-sections are well known so that,
at least at the present time, the intensity of intramolecular excitations is
more easily calculated than in IETS [3]. In addition, elastic and quasi-
elastic neutron scattering can be performed on the same samples used in the
inelastic experiments to study the structure and the diffusive motion of
molecules in the film.

1.2 Brief History of Neutron Scattering from Adsorbed Films

The first experiments applying neutron scattering to surface studies were
performed over a decade ago on hydrogenous films [4,6]. Hydrogen is particularly suitable as an adsorbate because it has the largest incoherent cross-
section (~80 barns) of any element. Unfortunately, due in part to the
heterogeneity of the substrates employed (charcoals and zeolites), well-
defined excitations such as found in bulk solids were not observed in

the early experiments on adsorbed films. However, significant progress was achieved in a series of experiments at Brookhaven beginning in 1973. These experiments involved both elastic and inelastic neutron scattering from films having a large coherent cross-section adsorbed on a high-homogeneity graphite substrate known at Grafoil. The structures of nitrogen [7], argon [8], and helium [9] films adsorbed on Grafoil were studied by elastic neutron scattering. Inelastic scattering was used to study the propagation of phonons in Ar^{36} monolayers [8] and rotons in thin helium films [9].

During this time, a number of neutron experiments on adsorbed films have been conducted at reactors in Europe. The structure of H_2, D_2, and O_2 adsorbed on Grafoil [10,11] and the dynamics of H_2 and D_2 on Grafoil [10] and alumina [12] were investigated at the Danish reactor, Risø. At the English reactor, Harwell, inelastic experiments on ethylene adsorbed on zeolites [13] and H_2 adsorbed on platinum [14] have been performed. Hydrogen chemisorbed on Rainey-nickel has been studied at Jülich [15] and both methane on graphite and neopentane on TiO_2 have been investigated at Grenoble [16]. Thus despite the intensity problem inherent in neutron scattering from adsorbed films, the technique has begun to be applied to a wide variety of films and substrates.

1.3 Motivation for the Present Work

Of particular interest in the inelastic neutron experiments on hydrogenous films has been the search for surface vibratory modes associated with the loss of translational and rotational degrees of freedom of the adsorbed molecule. One of our aims in initiating experiments was to return to the study of hydrogenous films to determine whether well-defined excitations such as surface vibratory modes and intramolecular vibrations could be observed in films adsorbed on high-homogeneity substrates. We also wished to address questions which previous experiments had not dealt with such as the application of neutron scattering to the determination of molecular orientation in adsorbed films and the determination of the location and strength of the bonds between the adsorbed molecule and the substrate.

2. Example of the Application of Neutron Scattering Techniques to the Study of Adsorbed Films: the Structure and Dynamics of Butane Adsorbed on Graphite

2.1 Experimental Considerations

To motivate the choice of butane as the adsorbate, we begin by reviewing the structure of the molecule. As shown In Fig. 1(a), butane $(CH_3(CH_2)_2CH_3)$ consists of hydrogen atoms tetrahedrally bonded to a skeleton of four coplanar carbon atoms. Associated with the terminal carbon atoms are CH_3 or methyl groups while the two interior carbon atoms are bonded to two hydrogen atoms forming CH_2 or methylene groups. One can see in Fig. 1(b) that there are symmetrically located planes of hydrogen atoms on each side of the carbon skeleton.

We shall be interested in only the lowest lying internal vibrations of the molecule which involve the torsional motion of the end methyl groups and of the two CH_2 groups with respect to each other. The torsional modes of short-chain paraffin molecules were first observed in neutron scattering experiments by BRUGGER and co-workers at the Idaho MTR reactor over ten years ago. In Fig. 2 we show their data for bulk solid butane at 106 K obtained on a time-of-flight spectrometer [17]. The observed dynamical structure

Butane $CH_3(CH_2)_2CH_3$

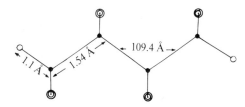

O Hydrogen

● Carbon

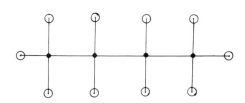

Fig.1 Structure of the butane molecule (C_{2h} symmetry). (a) The four carbon atoms (●) and one hydrogen atom from each CH_3 group are coplanar. The concentric circles (◎) represent the remaining hydrogen atoms located in two parallel planes one on each side of the carbon skeleton. (b) View showing the planar structure of the molecule

factor $S(Q,\Delta E)$ is proportional to the number of neutrons scattered with a particular energy loss ($\Delta E < 0$) or energy gain ($\Delta E > 0$) at a scattering angle of 4.1°. The energy-loss spectrum contains a well-resolved doublet at ~30 meV corresponding to the two methyl torsional modes and a smaller peak at ~20 meV corresponding to the CH_2-CH_2 torsion. There are several features of this spectrum which suggest butane as a suitable candidate for adsorption studies. The spectrum is not complicated, containing only three torsional modes of relatively high intensity which are well separated from a host of higher lying stretching and bending modes. Of all the intramolecular excitations we expect these modes to be perturbed most strongly upon adsorption, since they involve large amplitude motion of the hydrogen atoms.

Before performing a scattering experiment on a film, we first measure the adsorption isotherm consisting of the adsorbed volume of gas plotted against the vapor pressure in the cell. The isotherm for butane adsorbed on the powdered graphite substrate at T = 273 K appears in Fig. 3. The knee in the isotherm indicates monolayer completion at 0.55 liters (STP). By measuring the adsorption isotherm for nitrogen at T = 77K in the same cell and knowing the area occupied by a nitrogen molecule on graphite [7], we can also estimate the areal density of adsorbed butane and the surface area of the substrate. We estimate that the butane molecule occupies an area about 2.5 times that of nitrogen or ~40 $Å^2$. Such a large area per molecule provides the first evidence in our experiment that the molecule is adsorbed on its side rather than standing on end.

The substrate that we have used is a graphitized carbon powder known as Carbopack B [18]. The nitrogen isotherm gives a surface area of ~80 m^2/g which is almost three times that of the Grafoil substrate described above. Carbopack B is estimated to have a particle size of ~50 Å and the particles are assumed to be randomly oriented [2]. We selected it over the Grafoil

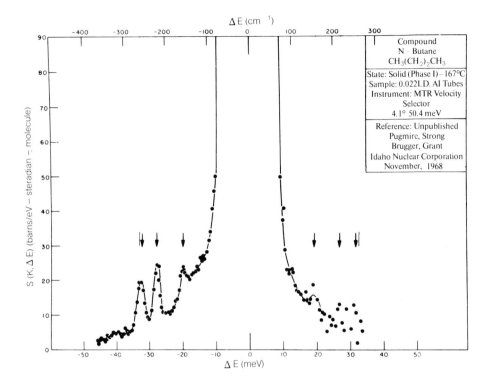

Compound
N – Butane
$CH_3(CH_2)_2CH_3$

State: Solid (Phase I) – 167°C
Sample: 0.022 I.D. Al Tubes
Instrument: MTR Velocity
Selector
4.1° 50.4 meV

Reference: Unpublished
Pugmire, Strong
Brugger, Grant
Idaho Nuclear Corporation
November, 1968

Fig.2 Time-of-flight spectrum for bulk solid butane at 106 K obtained at the Idaho MTR facility by BRUGGER and co-workers [17]

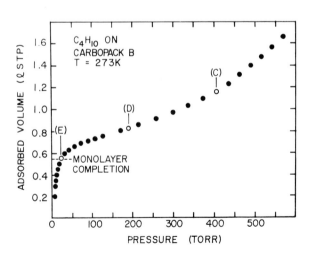

Fig.3 Adsorption iso-therm of butane on Carbopack B at 273 K

174

substrate discussed above for its larger surface area, high homeogeneity, and because the molecular excitations of interest could be studied without a preferred orientation in the substrate.

Before presenting the inelastic spectra, we briefly describe the time-of-flight spectrometer at the University of Missouri Research Reactor Facility on which these experiments were performed. The principle of operation is illustrated in Fig. 4. Neutrons originating in the reactor core pass through

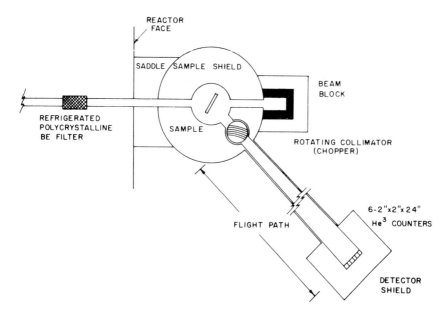

Fig.4 Schematic diagram of the time-of-flight spectrometer at the University of Missouri Research Reactor

a refrigerated beryllium filter which rejects all fast neutrons with wavelength less than 4 Å. The remaining low-energy neutrons ($\lesssim 5$ meV) are incident upon the sample. Those neutrons which gain energy by annihilating a phonon in the film are detected. Their energy is analyzed by a time-of-flight technique in which a chopper creates a burst of neutrons whose flight time to the detector bank 5.8 meters away can be measured. The detector is set at the smallest available scattering angle, 27.4°, in order to minimize rotational and translational (Doppler) broadening of the vibrational bands. Since the observed intensity of the modes depends upon their thermal population, the useful energy range of the spectrometer at 77 K is from about 30 cm^{-1} to 270 cm^{-1} over which the momentum transfer Q varies from 1 Å^{-1} to 3.1 Å^{-1}. The energy resolution of the spectrometer is about 10 cm^{-1} in this range. The fact that the energy-gain spectra are not taken at fixed momentum transfer is of little concern since the excitations observed are usually dispersionless.

2.2 Inelastic Spectra of the Butane Films

One of the advantages of the neutron technique is that one can obtain spectra as a function of film thickness from submonolayer coverages all the way up to the bulk limit. The inelastic spectra for butane adsorbed on Carbopack B at 77 K are shown in Fig. 5. Spectrum (A) is obtained from the bulk polycrystalline solid. Because our spectrometer operates in a neutron-energy-gain

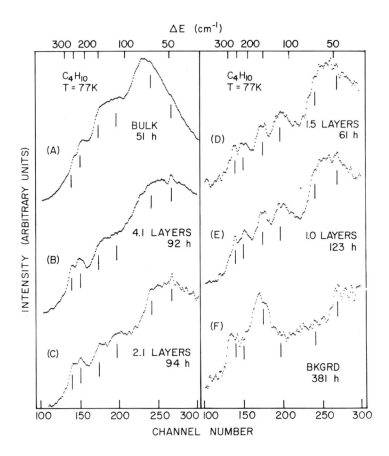

Fig.5 Inelastic spectra of butane adsorbed on Carbopack B at 77 K vs film thickness. Spectra of the bulk solid and the background with no butane in the cell are included for comparison. The counting times shown reflect the relative intensity of the spectra

mode at a larger scattering angle, the intramolecular modes are not as well resolved as in the Idaho data [17]. The intramolecular torsions appear as shoulders on a broad band attributed to intermolecular excitations. However, as the coverage is decreased from four layers to monolayer completion (spectra (B) through (E)), these shoulders evolve into a series of reproducible peaks.

Probably the most interesting feature is simply the number of peaks observed. In the monolayer spectrum there are four well-defined peaks and a broader band centered at 50 cm^{-1} compared with only three peaks observed in the bulk spectrum (see Fig. 2). The presence of these additional peaks suggests the existence of surface vibratory modes in which the adsorbed molecule vibrates against the substrate. Our aim is to try to understand the origin of these new vibrations by determining their eigenvectors, and, also, to investigate whether the monolayer spectrum contains information concerning the orientation and bonding of the adsorbed molecules.

To give some feeling for the scattered intensities involved, the monolayer spectrum represents a run of 123 h on the spectrometer. Therefore, data collection times are very long compared with those in IETS. The inelastic scattering from the substrate alone is shown in Fig. 5(F). It was necessary to count about three times longer for the background spectrum than for the monolayer spectrum to obtain comparable statistics. Hence at coverages above a monolayer the substrate does not contribute significantly to the spectra.

To facilitate comparison of the bulk and film spectra, we have included a schematic energy level diagram in Fig. 6. The energy levels for the three intramolecular torsions in bulk butane have been estimated from the Idaho data [17]. The width of the levels in the liquid state should not be interpreted quantitatively, but rather indicates qualitatively the effect of Doppler broadening. The vibrational states in the liquid phase are of interest to us because they give the best approximation to those of the free molecule, unperturbed by the presence of crystalline fields. In passing from the liquid to the solid state, the width of the excitations decreases and the energy of the lowest level, the CH_2-CH_2 torsion increases slightly. The intramolecular excitations appear at the same energy in the bulk solid and monolayer film but, as we have seen, new film-substrate modes are introduced. The larger width found for the film excitations probably results from the larger momentum transfers used in our measurements of the film spectrum.

2.3 Modeling of the Monolayer Spectrum

We have found the principal features of the monolayer spectrum can be reproduced by assuming that the molecule is adsorbed lying on its side with the plane of the carbon skeleton parallel to the graphite basal-plane surfaces. This orientation is depicted schematically in the inset to Fig. 7. For clarity, only the four coplanar hydrogen atoms which lie nearest to the surface (open circles) have been included. It is a remarkable feature of this orientation that the distance between hydrogen atoms in two adjacent CH_2 groups just calipers the carbon hexagon in the graphite basal plane. This geometrical coincidence suggested to us that the two methylene hydrogens are situated over the electron π-lobes projecting normal to the graphite layers. However, it is also possible that the molecule is translated half a lattice spacing so that the methylene hydrogens occupy the center of the carbon hexagons. Since in either case the two methylene hydrogens lie in symmetry-equivalent potential wells, we were lead to introducing the same force constant between each CH_2 hydrogen and the substrate. We also introduced a second force constant binding the co-planar methylhydrogens to the substrate. All force constants are directed along the surface normal. We assume that the molecule is only weakly perturbed upon adsorption so that the intramolecular force constants and atomic positions of the free molecule can be used [19].

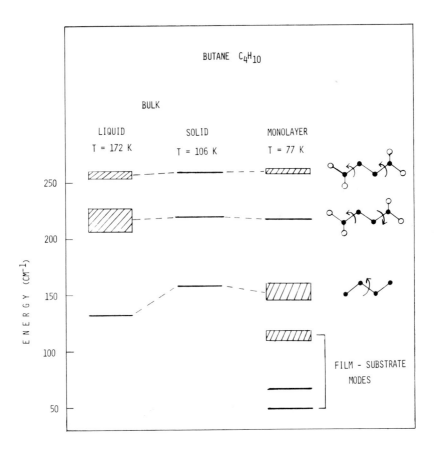

<u>Fig.6</u> Energy level diagram comparing the excitation spectra of bulk liquid and solid butane [17] with the monolayer spectrum (this work). The widths of the levels are not to be interpreted quantitatively. Diagrams at right represent the eigenvectors of the three intramolecular torsional modes

The normal mode problem is then solved to give the vibrational frequencies and eigenvectors of the adsorbed molecule. From these the relative intensity of the excitations can be calculated in the one-phonon approximation [1,2]. The calculated spectrum is shown below the experimental one in Fig. 7. The observed spectrum is for adsorbed monolayer (Fig. 6(E)) but with the background (Fig. 6(F)) subtracted. One can see that there is reasonable agreement between the calculated spectrum both with respect to the calculated energies and the relative intensities. The agreement worsens below 80 cm^{-1} where a broad band is observed. We believe that the two surface vibratory modes which are calculated to appear within this band are unresolved as a result of the butane-butane interaction which is neglected in the model.

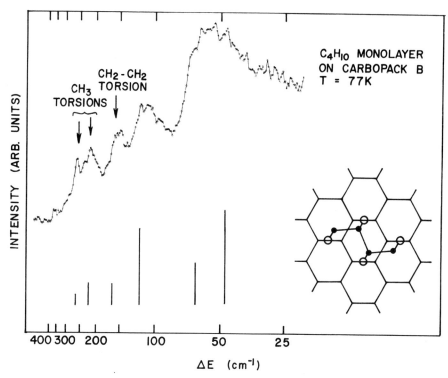

Fig.7 Comparison of the monolayer and calculated butane spectra. The background has been subtracted from the observed spectrum, and the error bars represent the statistical uncertainty. The arrows indicate the energies of the three lowest lying intramolecular modes of the bulk solid (Fig. 2). The calculated spectrum is for the two-parameter model described in the text. The inset shows the proposed orientation of butane with respect to the graphite basal plane. Only the four coplanar hydrogen atoms (O) nearest to the surface have been included for clarity

In this model three surface vibratory modes are introduced. The intense peak observed at 112 cm^{-1} is predicted to be a rocking mode about the symmetry axis of the molecule in the chain direction. The two unresolved surface vibratory modes within the band centered at 50 cm^{-1} correspond to a rocking motion about an axis orthogonal to the chain direction in the plane of the carbon skeleton and a uniform oscillatory motion of the entire molecule normal to the surface. The uniform oscillatory mode is less intense than the orthogonal rocking mode.

Although this model gives a good fit to the observed spectrum, it is necessary to check its uniqueness by examining other orientations and surface bonds of the adsorbed molecule. For example, we have verified that the agreement with the observed spectrum is not as good when the molecule is oriented on its side but with the carbon skeleton perpendicular to the surface [2]. Also, we have found for the parallel orientation considered above but with the bond to the surface localized near the butane carbon atoms neither the

179

frequencies nor the intensities of the surface vibratory modes are reproduced as well. The senstivity of the model spectra to the molecular orientation and bonding suggests that the neutron spectra can be a valuable aid in determining the local environment of an adsorbed molecule.

In order to fit the observed vibrational frequencies, we found it necessary to assume that the force constant binding the methyl hydrogens to the surface was about a factor of three smaller than for the CH_2 hydrogens. The bond between the CH_2 hydrogens and the graphite is itself weak, the force constant being only ~1% of that used for the C-H stretch in butane. This result is consistent with our initial assumption that the molecule is only weakly perturbed upon adsorption.

It is clear that a complete theory of the dynamics of adsorbed butane should include the interaction between molecules on the surface. Nevertheless, on the basis of our results for butane on graphite, it appears possible to deduce both structural and dynamical properties of the film from the inelastic spectra while neglecting the coupling between the molecules. We expect this approximation to be valid in other cases where the interaction between the adsorbate and the surface is sufficiently strong.

2.4 Elastic Neutron Diffraction from Adsorbed Butane

Having deduced a molecular orientation for butane on graphite from the inelastic spectra, we then performed elastic neutron diffraction to determine whether the film had long-range order and, if so, whether the lattice constant of the film was consistent with this orientation. The diffraction pattern from a 1.5-layer film at 81 K is shown in Fig. 8. The elastic scattering from the substrate has been subtracted and the neutron wavelength

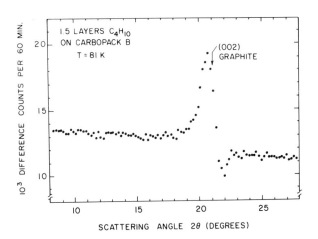

Fig.8 Elastic neutron diffraction from a 1.5-layer butane film at 81 K. The neutron wavelength is 1.27 Å. Background scattering from the substrate has been subtracted

is 1.27 Å. Somewhat surprisingly, we observe no Bragg peaks corresponding to an ordered two-dimensional phase of the film. However, at this point we cannot say whether the film is disordered or whether it is attempting to order but with a unit cell of dimensions comparable to a graphite particle size (~50 Å). Measurements using a grafoil substrate having a larger particle size (~100 Å) and deuterated films having a larger coherent cross-section should help to resolve this question.

Despite the absence of Bragg peaks, the diffraction pattern does contain structural information. There is a prominent modulation of the 002 graphite reflection due to the presence of the butane overlayer. The modulation results from constructive interference between neutrons scattered from the adsorbed monolayer and the graphite basal planes. Therefore, we can infer that most of the adsorption takes place on the exposed basal-plane surfaces of the graphite particles. Furthermore, the angular dependence of the modulation--the fact that the intensity peaks at an angle slightly less than the 002 graphite peak and passes through a minimum at a somewhat larger angle--is sensitive to the molecular orientation within the overlayer. Our preliminary analysis [20] suggests that the orientation of the butane molecule deduced from the inelastic spectra is consistent with the observed 002 modulation.

3. Future Experiments and the Interaction of IETS with Neutron Molecular Spectroscopy

We are presently extending our neutron scattering experiments to other short-chain hydrocarbons adsorbed on a variety of substrates. Preliminary measurements have been completed on ethane (CH_3CH_3) and propane ($CH_3CH_2CH_3$) adsorbed on graphite. Although one might expect a more easily interpretable results from the shortest chain, ethane, we find that the observed spectrum does not contain well-resolved surface vibratory modes. As in the case of butane, the methyl groups do not appear to interact as strongly with the graphite as do the CH_2 groups so that the intermolcular forces become relatively more important. Our tentative conclusion is that the effect of the inter-molecular coupling is to spread the surface vibratory modes into a broad band. The situation improves in propane, having one CH_2 group, where a surface vibratory mode appears as a high-energy shoulder on a broad, low-frequency band. Our model calculations suggest that this shoulder is the rocking mode about the chain axis which develops into a well-resolved peak at 112 cm^{-1} in the case of butane. Because of the broader features in the inelastic spectra of the shorter hydrocarbon films, it is difficult to obtain the molecular orientation and surface bonds from these experiments alone. Thus we believe it will be useful to investigate the 002 modulation in deuterated films of ethane and propane as well.

The application of neutron molecular spectroscopy to adsorbed films is not limited to chemically inert substrates such as graphite. The technique can also be applied to hydrocarbons chemisorbed on catalytically active transition metal powders such as nickel, platinum, and palladium. These powders do not adsorb neutrons strongly and are commercially available as catalysts with large surface areas. Experiments with hydrogen chemisorbed on nickel [14] and platinum [15] have found evidence for a large amplitude vibration of the hydrogen molecule against the metal substrate.

As far as interaction with IETS is concerned, neutron experiments with powdered Al_2O_3 substrates would probably be of the most immediate value.

Using an approach similar to that which we have described for butane on graphite, one would hope to answer such questions as the molecular orientation of hydrocarbons adsorbed on Al_2O_3 surfaces and whether the adsorbed molecules penetrate the bulk through pores. Since IETS spectra generally do not show the frequencies of stretching and bending vibrational modes of hydrocarbons shifted from those of the free molecule, it would be of interest to determine whether such is also the case for low-frequency torsional modes which can be investigated by neutron scattering.

We anticipate that there will be more interaction between the neutron and electron surface spectroscopies in the future as pulsed neutron sources offering higher fluxes of neutrons at energies above 0.1 V are developed. The WNR pulsed neutron source is now operational at Los Alamos Scientific Laboratory and the Zing-P' source will soon be operational at Argonne National Laboratory. The pulsed sources promise to extend inelastic neutron scattering to the energy range of the electron and optical spectroscopies while retaining some unique features of the neutron as a surface probe. We conclude by briefly summarizing these capabilities: The neutron technique utilizes well-characterized powdered substrates whose adsorption surfaces can be determined by the overlayer modulation of a substrate Bragg peak. Neutron experiments can be done at high ambient pressures and temperatures such as would be found in a catalytic reactor. Finally, inelastic and elastic scattering can be performed on the same film so that both the dynamics and the structure of the absorbate can be investigated.

[1]Work supported by the University of Missouri Research Council, the University of Missouri Research Reactor Facility, and the Research Corporation

Acknowledgement

The author is deeply grateful for the collaboration of H.R. Danner, Y.P. Sharma, H.L. McMurry, and R.M. Brugger in these experiments.

References

1 H. Taub, H.R. Danner, Y.P. Sharma, H.L. McMurry, R.M. Brugger: Phys. Rev. Lett. 39, 215 (1977)
2 H. Taub, H.R. Danner, Y.P. Sharma, H.L. McMurry, R.M. Brugger: Proceedings of the IV Rolla Conference on Surface Properties of Materials, Aug. 1-4, 1977, Rolla, Mo. and to be published in Surf. Sci.
3 B. Hudson, A. Warshel, R.G. Gordon: J. Chem. Phys. 61, 2929 (1974)
4 H. Boutin, S. Yip: Molecular Spectroscopy with Neutrons (M.I.T. Press, Cambridge, 1968) and reference cited therein
5 S. Todireanu: Nuovo Cimento Suppl. 5, 543 (1967)
6 J.W. White: in Proceedings of the Fifth IAEA Symposium on Neutron Inelastic Scattering, Grenoble, France, 1972 (International Atomic Energy Agency, Vienna, 1972), p. 315

7 J.K. Kjems, L. Passell, H. Taub, J.G. Dash, A.D. Novaco: Phys. Rev. B 13, 1446 (1976)
8 H. Taub, L. Passell, J.K. Kjems, K. Carneiro, J.P. McTague, J.G. Dash: Phys. Rev. Lett. 34, 654 (1975); and to be published in Phys. Rev. B
9 K. Carneiro, W.D. Ellenson, L. Passell, J.P. McTague, H. Taub: Phys. Rev. Lett. 37, 1695 (1976)
10 M. Nielsen, W.D. Ellenson: in Proceedings of the Fourteenth International Conference on Low Temperature Physics, Otaniemi, Finland, 1975, edited by M. Krusius and M. Vuorio (North-Holland, Amsterdam, 1975), Vol. 4, p. 437
11 J.P. McTague, M. Nielsen: Phys. Rev. Lett. 37, 596 (1976)
12 I.F. Silvera, M. Nielsen: Phys. Rev. Lett. 37, 1275 (1976)
13 J. Howard, T.C. Waddington, C.J. Wright: J. Chem. Soc. D 789, 775 (1975); and to be published
14 J. Howard, T.C. Waddinton, C.J. Wright: J. Chem. Phys. 64, 3897 (1976)
15 R. Stockmeyer, H.M. Conrad, A. Renouprez, P. Foulilloux: Surf. Sci. 49, 549 (1975)
16 J. White, R.K. Thomas, T. Trewern, I. Marlow, A. Miller: Proceedings of the IV Rolla Conference on Surface Properties of Materials, Aug. 1-4, 1977, Rolla, Mo. and to be published in Surf. Sci.
17 K.A. Strong: Catalogue of Neutron Molecular Spectra, AEC Report IN-1237, Idaho Nuclear Corp.
18 Supelco, Inc., Bellefonte, Pa. 16823. Carbopack B is similar to Graphon previously manufactured by the Cabot Corp., Billerica, Mass. 01821
19 K.W. Logan, H.R. Danner, J.D. Gault, H. Kim: J. Chem. Phys. 59, 2305 (1973)
20 H. Taub, H.R. Danner, L. Passell: to be published

VI. New Applications of IETS

Study of Supported Catalyst Particles by Tunneling Spectroscopy [1]

P.K. Hansma [2]

Department of Physics, University of California
Santa Barbara, CA 93106, USA

ABSTRACT

It is possible to evaporate a small quantity of metal onto the oxidized a-luminum strip of a tunnel junction and then study the adsorption of gases on the resultant supported metal particles. This talk will focus on the adsorption of CO on supported Rh particles. It will also include recent data on the reaction of H_2 with this adsorbed CO. Research on this type of system may help in understanding Fischer-Tropsch synthesis and other important catalytic reactions.

Most inelastic electron tunneling spectroscopy experiments have studied molecules adsorbed onto oxidized aluminum [1]. Today I would like to show you that it is also possible to study molecules adsorbed on small metal particles supported on this oxidized aluminum.

I would like to begin by describing how the junctions are made. First, an aluminum electrode is evaporated and oxidized. Then the oxide is cleaned using the method of ADLER and co-workers: exposing the oxide to an argon glow discharge [2]. I use approximately 300 milliamps for 3 minutes at 50 μ pressure. Next I evaporate a small amount of metal onto the cleaned oxide. In the experiments I'll describe today, that metal was rhodium, and by a 'small amount' I mean somewhere between 1 and 4 angstroms average coverage as measured with a quartz crystal thickness monitor.

I don't know how the rhodium is dispersed on the oxide. Specifically, I don't know how much agglomeration occurs. I would assume, however, that the agglomeration is not extensive, since the substrate temperature is only about 0.15 of the melting temperature of Rh [3]. But, that is speculation at this point. I am beginning a cooperative experiment with JIM SCHWARZ and TERRY BAKER at Exxon to do high resolution electron microscopy [4] on these samples to try to find out what the particle size distribution really is.

At any rate, we now have a little rhodium on top of aluminum oxide. This rhodium metal can be doped. In the experiments I will describe today, it has been doped by exposure to approximately 10^{-5} torr of CO for 100 seconds. Since rhodium metal is very reactive, it is important to minimize the time and pressure between completion of the rhodium evaporation and doping. In practice I often pump down to approximately 10^{-7} torr and then backfill the chamber with 10^{-5} torr of carbon monoxide during the rhodium evaporation. At 10^{-5} torr the mean free path is such that there is a small chance that a rhodium atom is going to hit a CO molecule on the way to the substrate, but once it hits the substrate, CO dominates the residual gases in the chamber.

Finally, I complete the junction with the usual top lead electrode.

The first slide shows some results that I obtained in collaboration with BILL KASKA and RICK LAINE who were both in the Chemistry Department of UC Santa Barbara [5]. The lower trace is a spectrum with no rhodium evaporated into the aluminum oxide. This is just the usual background trace with modes due to aluminum phonons, aluminum oxide phonons, the second harmonic of aluminum oxide phonons, and OH groups. In addition, these early samples had some hydrocarbon impurities as evidenced by the CH stretch mode, but, as you will see, this is very much decreased in more recent samples. At any rate, the point is that there are no peaks characteristic of carbon monoxide. The next trace shows the results for $\frac{1}{2}$ Å of rhodium (approximately 1/6 of a monolayer if it were uniform). The most prominent new structure is a little peak near 400 cm^{-1}, a region that metal organic chemists know is characteristic of the vibrations of CO coordinated to a transition metal [6]. As the rhodium coverage is increased, this peak grows and other peaks appear. The most interesting thing to me is that as coverage increases you don't get just an increase in peak intensity; rather you get growth of different peaks. We believe that the peak at 408 cm^{-1} and the higher energy peak in the CO stretching region at 1935 correspond to carbon monoxide that is linearly bound to a single rhodium atom, and that the peak near 580 cm^{-1} and the lower energy peak in the CO stretching region at 1730 cm^{-1} correspond to a bridged species. These assignments are tentative at the moment. RICHARD KROEKER is beginning experiments with isotopically labeled carbon monoxide to unambiguously assign the peaks.

The next slide shows a differential tunneling spectrum obtained with the bridge [7] that I mentioned at the first of my talk on Wednesday. This slide also shows the single junction tunneling spectrum of a junction with rhodium and of the control junction without rhodium. Both junctions were made on the same aluminum strip, both were exposed to CO, both were completed with an evaporated lead electrode at the same time. The only difference is that one had a little bit of rhodium evaporated before the CO exposure, and the other didn't. One of the first things you notice about the differential tunneling spectrum is that it doesn't help you nearly as much in this case as it did for the liquid doped case I showed on Wednesday. My speculation is that the evaporated rhodium metal modified the tunneling barrier and perhaps the superconductivity of the lead. At any rate, though the differential tunneling bridge doesn't help as much, it does help. In particular, even at this low coverage you can see both of the peaks in the CO stretching region that were only clearly resolved at higher coverages in the previous slide. Also, you get a better idea of what the shape of the structure actually is.

At this point I feel it is appropriate to confess that most of the experiments I have attempted on supported metal particles have failed. Figures 1 and 2 were taken almost a year apart. In between I tried to absorb CO on a variety of other supported metals and to absorb a variety of other compounds on supported rhodium. Almost all of these experiments failed. As exceptions I was able to get several good spectra for carbon monoxide adsorbed on nickel and some fairly good spectra for methanol and etahanol adsorbed on rhodium. But there are a wide variety of things that can go wrong

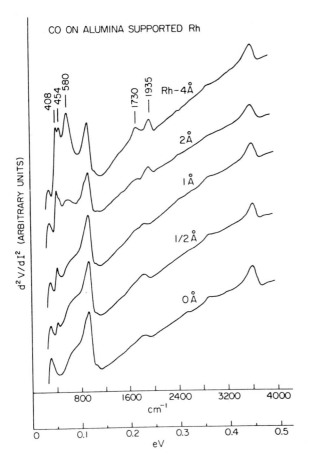

Fig.1 Tunneling spectra of CO adsorbed on various thicknesses of Rh evaporated on alumina. Note that new peaks appear as the thickness of Rh increases implying new types of bonding. For all these spectra the CO exposure was $\approx 10^3$ Langmuir (1 Langmuir = 10^{-6} torr sec). They were taken with a 2 mV modulation voltage at 4.2°K

in this type of experiments: the zero bias anomolies you heard about this morning and the lack of adsorption of compounds that should have adsorbed were my most common problems.

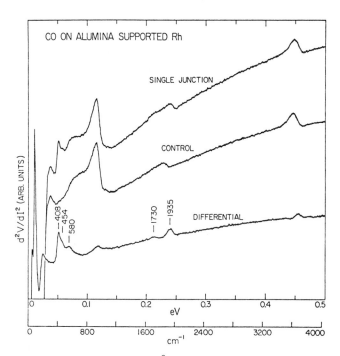

Fig.2 Approximately 1.5 Å of Rh were deposited onto one oxidized aluminum
strip. Another, on the same substrate, received no Rh. Both were exposed
to ≈ 10^3 Langmuir of CO. Both were then covered with an evaporated Pb top
electrode. The upper trace is the single junction tunneling spectrum of the
junction with Rh, the middle trace is the spectrum of the one without. The
lower trace is the difference spectrum. The spectra were taken with a 2 mV
modulation voltage at 4.2°K

There are, however, a wide variety of tricks that can be employed in the
future. For example, heated substrates may improve reactivity, cooled substrates may hold on weakly adsorbed species, and the evaporated transition
metal or top electrode metal can be thermalized with an argon gas carrier to
reduce their kinetic energy. Hopefully this type of trick will help with
these relatively difficult experiments.

But I certainly don't want to end this talk on a discouraging note, because I am very excited. About a month ago I talked to BOB JAKLEVIC on the
phone about the new results that he will talk to you about next. I figure
that if he can get all of those big organic molecules down through the top
metal electrode, I should be able to get hydrogen through the top metal
electrode and possibly react it with the CO adsorbed on rhodium. The importance of this possibility is that it relates to making hydrocarbons from
the products of coal gasification: CO and H_2. Furthermore, there is almost
no spectroscopic information on what the reaction intermediates are for this
important reaction.

Thus the plan was to complete a junction with CO adsorbed on rhodium and
expose it to high pressure hydrogen at various temperatures and then study the
resultant spectra. Figure 3 shows the results on a junction that was cycled

189

five times by warming it with dry compressed air as it was taken from the dewar. The bottom spectrum is before hydrogen treatment. It looks very similar to the 2Å trace of Fig. 1. The upper spectra are labeled by the maximum temperature to which the junction was heated in a cell with approximately 100 atmospheres of high purity hydrogen. Note that the bands due to the adsorbed carbon monoxide decrease in intensity and that new bands in the CH deformation and stretch region appear. At the highest temperature even these hydrocarbon bands begin to decrease.

The spectra of Fig. 3 are single junction spectra. Figure 4 shows differential tunneling spectra for a sample with a comparable amount of rhodium

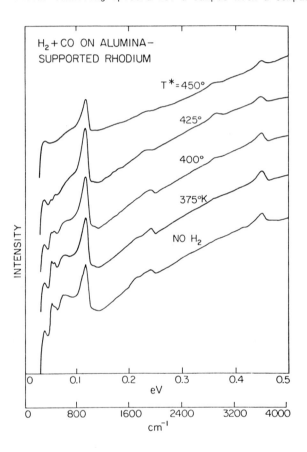

Fig.3 The lowest curve is the tunneling spectrum of a junction with ≈ 2 Å of Rh that was exposed to ≈ 10^3 Langmuir of CO. The other curve show its tunneling spectrum after exposure to ≈ 100 atm of high purity H_2 at various temperatures. The temperature of the sample was approximately 2°K lower than the temperature of the high pressure cell, T*. The spectra were taken with a 2 mV modulation voltage at 4.2°K

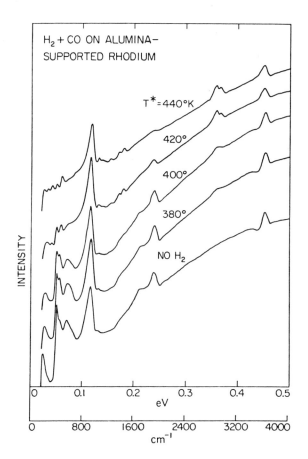

Fig.4 The lowest curve is the differential tunneling spectrum of a junction with ≈ 2Å of Rh that was exposed to ≈ 10³ Langmuir of CO. The other curves show its differential tunneling spectrum after exposure to ≈ 100 atm of high purity H₂ at various temperatures. The temperature of the sample was approximately 2°K lower than the temperature of the high pressure cell, T*. The spectra were taken with a 2 mV modulation voltage at 4.2°K

(approximately 2Å). Note that again the differential tunneling spectrum does help distinguish the structure due to adsorbed species from the background structure although it does not completely eliminate the background structure. Again, as in Fig. 3, the general feature of this figure is that as the temperature at which the junction is exposed to hydrogen is increased, the bands due to adsorbed carbon monoxide decrease and new bands, due presumably to adsorbed hydrocarbon, increase. Specific features of these traces, in particular the series of small peaks below 400 cm⁻¹ in the trace labeled 440°K, do not always reproduce. It may be that the exact reaction product depends on variables that we do not yet have under control. We are continuing our experiments.

On each of the slides with the junctions used for Figs. 3 and 4 was a control junction that was treated identically in every way except that it did not have evaporated rhodium. The single junction spectra for this control junction (analogous to Fig. 3) showed no adsorbed CO and no hydrocarbon formation. (The control junction spectrum is, of course, subtracted from the rhodium doped junction in Fig. 4.) The lack of hydrocarbon in the control junction spectra is the only reason that I am confident enough in this data to be willing to show it to you today. It says to me that if the results are due to the diffusion of a hydrocarbon impurity into the junction area, it is at least a very choosey hydrocarbon: it will not appear in a junction without rhodium metal, and in a junction with rhodium metal it displaces adsorbed CO. Fortunately, I believe the more exciting possibility is more probable: hydrogen (either from the gas or dissolved in the metal electrode, or even from OH groups on the surface) is reacting with the adsorbed CO to form hydrocarbons.

RICHARD KROEKER has joined this project and is working with isotropically labeled carbon monoxide: specifically, $C^{13}O$ and CO^{18}. It is our hope that the small isotopic shifts in bands will allow us to: (1) determine the structure and force constants of the adsorbed carbon monoxide species; (2) determine whether the hydrocarbon we see forming is a true reaction product as opposed to a diffusing impurity; (3) identify the adsorbed hydrocarbon. We believe that the identification of this hydrocarbon and the extension of this experiment to other reaction conditions and systems may help us to make an important contribution to Fischer-Tropsch chemistry.

[1]Work supported by the National Science Foundation
[2]Alfred P. Sloan Foundation Fellow (1975-1977)

References

1 A recent review is P.K. Hansma: Physics Reports 30C, 145 (1977)
2 R. Magno, J.G. Adler: Phys. Rev. B 13, 2262 (1976)
3 At low enough substrate temperatures surface diffusion is small and even monomers are stable to reevaporation
4 As done, for example, by E.B. Prestridge, D.J.C. Yates: Nature 234, 345 (1971)
5 P.K. Hansma, W.C. Kaska, R.M. Laine: J. Am. Chem. Soc. 98, 6064 (1976)
6 W.P. Griffith, A.J. Wickham: J. Chem. Soc. A, 834 (1969); L.M. Vallarino: Inorg. Chem. 4, 161 (1965)
7 S. Colley, P.K. Hansma: to appear in Rev. Sci. Inst., Sept. 1977

External Doping of Tunnel Junctions

R.C. Jaklevic and M.R. Gaerttner

Research Staff, Ford Motor Company
Dearborn, MI 48121, USA

ABSTRACT

Inelastic tunneling spectroscopy (IETS) with M-I-M diodes has been used to detect and study the vibrational modes of organic molecules contained in monolayer amounts inside the barrier layer. In previous experiments it has been necessary to introduce these molecules as an intermediate step in the fabrication process with an additional vacuum evaporation of the top metal electrode coming last. We have developed a method whereby the completed and undoped tunnel junctions can be fabricated in batches and stored in air for later doping with various organic molecules. The doping method is a vapor exposure technique in which the presence of water vapor is a condition for the infusion of molecules into the barrier region. A microscopic layer of adsorbed water apparently acts as a carrier for the introduction of molecules. Experiments with D_2O show that the OH can be exchanged with OD inside the barrier. A large number of polar organic molecules (e.g. formic acid, methyl alcohol, hydrogen cyanide) have been infused in monolayer coverages in a few minutes by this method. Deuteration experiments have verified this interpretation. The sensitivity and selectivity of this technique is illustrated by the ability to detect low levels (<10 ppm) of formic acid in singly distilled or deionized tap water. Some of the species studied (methyl alcohol, pyridine) could not be studied easily by previous IETS doping methods. Further study is in progress to understand the details of the doping mechanism and determine the size and types of detectable molecular species. This approach to IETS has been found easier and more flexible than before and is capable of high sensitivity and selectivity to certain types of molecules.

The experiments described here are the result of an attempt to see if there is a way of doping tunnel junctions after they are made, removed from the vacuum system and stored on the shelf. As it turns out, we have succeeded in doing this [1] and this paper will be a description of our experiments and some results obtained so far. Although there are many things that are still not understood, we have a number of important clues about the way the doping works and the kind of molecules which can be used.

The junctions are made in the conventional way by evaporation in a bell jar with an electronically pumped oil-free vacuum system. We start with the Aℓ layer about 700Å thickness and then grow a thin oxide layer by exposure either to O_2 at a few torr pressure or by exposure to an O_2 discharge for a few minutes. After growing the oxide, the cross strip (usually Pb) is evaporated to a thickness of about 700Å (2 ohms per square). The measured resistances vary from less than 0.1 ohms to over 100 ohms with gas discharge oxidation yielding the higher values. The junctions are stored in a dry box where we observe that their resistances and IETS spectra will remain unchanged for a period of weeks or months. For the case where the resistances are large enough (above 20 ohms) that their IETS spectra can be run, we find little or no residual hydrocarbon impurities, showing that the junctions are initially clean.

What will be described is an infusion technique whereby various organic molecules can be introduced into a completed tunnel junction. In previous methods of doping, the organic layer is introduced as a step during junction fabrication where the molecules are deposited (either from vapor or solution) and the final step of vacuum evaporation is necessary to complete the junction. In the infusion method Fig. 1, the completed junction is removed from storage and placed in a small chamber (humidor) in which there is a small amount of water which can have an organic chemical in it. Inside is a smaller beaker into which is placed the junction substrate and the container is covered by a closely fitting lid. The container is placed in a small tray of water (not shown) to keep the chamber at slightly lower temperature than ambient. Thus the junctions are exposed to nearly 100% relative humidity plus the vapor of the organic molecules. The temperature of the water is kept slightly below ambient in order to prevent dew from forming on the substrate and ruining the Pb film.

In the relative humidity range above about 50% a number of effects happen. First, the resistance of the junctions increase with time, rising from less than 0.1 ohms to 100 ohms or more over a period of time from ten minutes to one day. It is necessary to time the exposure and monitor the resistance in order to stop the process before the resistance goes over a few thousand ohms.

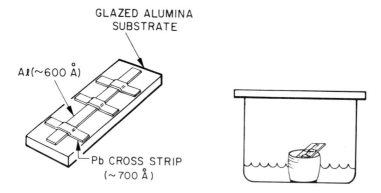

GLAZED ALUMINA
SUBSTRATE

Aℓ(~600 Å)

Pb CROSS STRIP
(~700 Å)

Fig.1   Schematic showing junction and basic geometry of the humidifier infusion chamber. The water can contain the organic molecules of interest or they may be deposited directly over the junction and substrate

The higher the humidity, the faster the rate of increase; for example at 50% relative humidity a day is required to reach several hundred ohms. If the junctions are stored in less than 30% relative humidity, the change in resistance, if any, is very slow. A second effect of exposure to humid vapor is the appearance of a certain amount of structural change of the Pb film with the most drastic effect occurring for higher humidities. The sheet resistance of the Pb rises as much as 10% while the degree of surface scattering of incident light (as observed by eye) noticeably increases. This indicates a probable increase in grain size of the Pb. For lower humidity exposures this effect becomes much less pronounced. There is no experimental reason that this structure change is essential to the success of the infusion technique since highly doped junctions have been produced with a minimum of Pb film change. Pb thicknesses from 200 Å to 1500 Å have all been employed with success and Pb films deposited on cooled substrates can also be doped by infusion.

During the period while the resistance increase occurs, we find experimentally that the IETS spectrum changes - both water and other types of molecular species can enter and leave the tunnel junction. The experimental evidence for this will be presented and an attempt will be made to explain some of the details. The first experiment, shown in Fig. 2 is for the case of H_2O infusion. Curve A is the spectrum obtained from a junction prepared by introducing D_2O into the vacuum system during oxidation. The OD stretching vibration is at about 327 mV. After infusion of H_2O in nearly 100% relative humidity for thirty minutes the spectrum of curve B is obtained. The OD line is nearly absent and the intensity of the OH increased. The reverse experiment was also performed. Curves C and D are samples prepared in the same batch. Curve C shows no OD initially present. After infusion of D_2O in nearly 100% relative humidity for forty minutes the spectrum of curve D is obtained. A strong OD line is now apparent and the OH line has become weak. These experiments show that water can go in and out of a tunnel junction by this technique.

Next we look at data for organic molecules [2]. The one studied most was

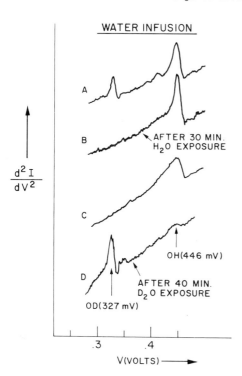

Fig.2 IETS spectra showing water infusion for a sample with initial OD vibration (curve A) and the same sample after infusion with H_2O (curve B). Also shown is a sample with no initial OD (curve C) and a sample from the same batch after infusion with D_2O (curve D). Exposure times were about thirty minutes

formic acid because of the ease with which it produced intense spectra. As previously mentioned, impurities are a constant problem, coming from outside sources such as dirty utensils, air and water. So it is useful to use deuterated chemicals whenever possible. Figure 3 shows an example of formic acid DCOOD spectra produced by both the usual way (vacuum deposition onto the freshly oxidized Aℓ) and be infusion from 500 ppm D_2O and 10 ppm D_2O for about five minutes. The lines at about 113 mV and 260 mV are characteristic of the CD deformation and stretching modes, respectively. Clearly the formic acid is being introduced into the tunnel junction and, for the bottom curve, the method is shown to have great sensitivity to very small amounts of formic acid. Although, for these spectra the intensities were not measured by an accurate method, a comparison with the Pb superconducting phonon spectrum indicates that both doping methods result in spectra of comparable intensity and correspond to roughly a monolayer of adsorbed impurities. We find that once a tunnel junction has been doped, it can be stored in a dry box for many weeks and the spectrum remains unchanged.

Formic acid spectra are hard to avoid in that it appears to be a common impurity encountered from uncontrolled outside sources. Figure 4 shows three spectra obtained from vacuum deposited normal formic acid and also from impurities infused from singly distilled water and from several days exposure to approximately 70% relative humidity room air. The lines at about 132 mV and 360 mV are characteristic CH deformation and stretching modes, respectively. It is not possible to estimate from these curves what impurity levels are involved but in no case was any detectable odor of formic acid

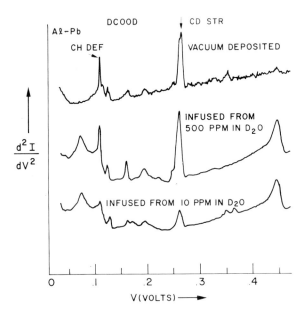

Fig.3 IETS spectra for DCOOD. The top spectra is obtained from a sample doped in the conventional manner before deposition of the Pb. The second two curves are from samples doped by infusion in a few minutes

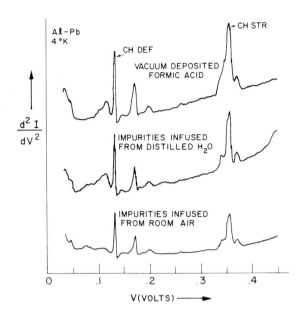

Fig.4 Comparison of IETS spectra obtained from normal formic acid (top curve) and impurities infused from singly distilled water and humid room air

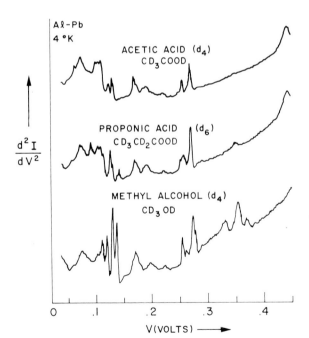

Fig. 5 IETS infusion spectra for deuterated acetic and propionic acids and methyl alcohol. Lines at 260 mV are characteristic of CD stretching vibrations

present. Formic acid has a vapor pressure very similar to water so it is not surprising that it might be residual impurity after distillation. We were able to obtain reasonably organic-free spectra with triply-distilled water.

In Fig. 5 is shown the spectra of acetic and propionic acids, both infused from solutions with D_2O. Clearly the presence of the CD stretching modes around 260 mV show the presence of the molecules inside the junction. Also shown is the example of methyl alcohol. Methyl alcohol is of special interest because it will not form a stable monolayer at room temperature on Aℓ oxide in a vacuum system.

In Fig. 6 are shown spectra for infused molecules of pyridine and both normal and deuterated acetone. Again, lines at about 260 mV are CD stretching modes. Pyridine is a ring compound, showing that even larger molecules can be infused. None of these spectra can be obtained by usual doping methods because they will not form stable room temperature monolayers. Figure 7 shows infusion spectra for both normal and deuterated forms of dimethyl-sulphoxide. Figure 8 shows the spectra of hydrogen cyanide, produced by including a small amount of KCN and H_2SO_4 in the water. The CN stretching

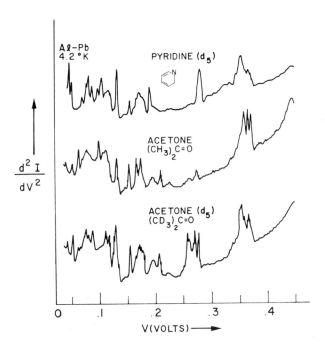

Fig.6 IETS infusion spectra for deuterated pyridine and acetone and normal acetone. Lines at 260 mV are characteristic of CD stretching vibrations

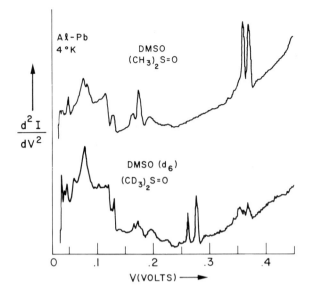

Fig.7 IETS infusion spectra for normal and deuterated dimethyl-sulfoxide. Both CH and CD stretching vibrations are seen at about 360 and 260 mV, respectively

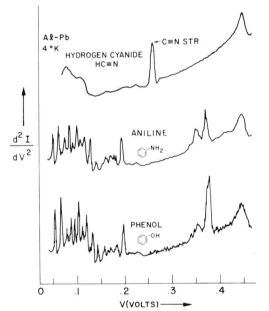

Fig.8 IETS infusion spectra for hydrogen cyanide, aniline, and phenol. The CN stretching mode at about 260 mV is characteristic while the structure around 198 and 410 mV for aniline is characteristic of NH stretching and bending vibrations. For aniline and phenol the aromatic CH stretching modes appear at about 372 mV

vibration at about 260 mV is characteristic. Also shown are two ring compounds, aniline and phenol. The former shows indications of the NH stretching vibrations around 420 mV and both show the characteristic aromatic CH stretching modes at about 372 mV.

To list the basic results of these experiments, first we have shown that water and a number of organic molecules including acids, bases, and ring compounds can be infused. Infusion time is from one minute up to one hour for coverages of about one monolayer. Some molecules, which can be introduced only with difficulty (acetone, pyridine, methyl alcohol) by vacuum adsorption methods can be infused simply. Also, we have found that pyridine can be removed by subsequent exposure to pure H_2O vapor. With the possible exception of acetone, all successful infusions have required the presense of water vapor. The Pb film thickness or smoothness does not strongly affect the rate of doping, rather molecular type and relative humidity are much more important. Sn electrodes have also been used successfully but so far have not been studied. The method can be quite sensitive, depending on molecular type (formic acid is most readily infused) so that there are selectivity and sensitivity properties which are potentially very useful. All of the molecules successfully infused so far are polar, i.e. are to some degree soluble in water.

All of the facts obtained so far lead us to a tentative picture for infusion which involves a two-step process. First, the molecules must be able to adsorb on the Aℓ and Pb oxide surfaces together with adsorbed water. The second step involves diffusion of the molecules into the region of the tunnel junction. Our evidence shows that the water is mobile and we believe that the water is necessary to promote diffusion of the polar molecules. Water adsorbed on the surface of aluminum oxide is a system familiar to the physical chemist [3]. Figure 9 shows a sketch of an adsorption isotherm in which it is seen that at roughly 50% relative humidity there is about one monolayer

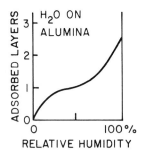 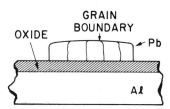

Fig.9 Schematic of an adsorption isotherm of water on γ alumina showing the monolayer type coverage near 50% relative humidity. Infusion works for humidities above 30%. The sketch shows the possible paths of diffusion along grain boundaries and into the tunneling interface

adsorbed and for higher humidity levels the average thickness is greater. These additional layers are bonded to the first layer with forces similar to those in liquid water. One might think of these surfaces as being covered by two dimensional liquid which act as a carrier for polar molecules. Similar adsorption isotherms exist for water on other materials. Of course, it must be remembered that Pb films, even when smooth, have grain boundaries extending right through them. Therefore, the possible paths of diffusing molecules will be along boundaries and subsequently along the Pb-Aℓ oxide interface.

200

There are a number of details which we must learn more about. What is the best material and method for making the metal films so that the infusion is rapid, reproducible, and versatile? Under what conditions can molecules be removed and others infused? What kinds and sizes of molecules can be used? How can the resistance rise on doping be related to the actual process of infusion itself? Control of unwanted impurities might become a problem for very critical experiments.

The infusion technique can be used to extend the usefulness of IETS. Adsorption and diffusion of two kinds of molecules might allow the observation of chemical reactions within the tunnel junction. Application to surface catalysis and adhesion problems are possible. The sensitivity and selectivity of the device indicates that it can be used for trace detection of certain impurities. Also it is a step in the direction of making IETS easier and more versatile in that it decouples the use of the device from the vacuum procedures necessary to fabricate it.

References

1   R.C. Jaklevic, M.R. Gaerttner: Applied Phys. Lett. 30, 646 (1977)
2   For reviews of IETS spectroscopy and analysis of these types of spectra
    see P.K. Hansma: Physics Reports 30C, 146 (1977) and R.G. Keil, T.P.
    Graham, K.P. Roenker: Appl. Spectroscopy 30, 1 (1976)
3   J.D. Carruthers et. al.: J. Colloid and Interface Science 36, 205 (1971)

Electronic Transitions
Observed by Inelastic Electron Tunneling Spectroscopy

S. de Cheveigné, J. Klein, and A. Léger

Groupe de Physique des Solides de L'E.N.S.[1]
Paris, France

ABSTRACT

We wish to discuss the application of IETS to the observation of <u>electronic</u>, rather than vibrational transitions. This extension of the method seemed interesting for several reasons.

First, one can expect electronic energy levels to be particularly sensitive to the binding of the molecule to the oxide layer, and to give useful information on absorption problems. What's more electronic transitions have much greater oscillator strengths than vibrational ones ($\simeq$ .1 for optically allowed transitions, against $= 10^{-5}$) they should be particularly easy to observe supposing that an "optical model" still holds.

This is very promising, but in practice a number of difficulties arise:

- The main one is that most electronic transitions are at energies much higher than the corresponding voltage limit on tunnel junctions (2-3 V). This will severely restrict our choice.
- Even within that limit, the inelastic component is a very small fraction of the total current, firstly because, above .5 V, the elastic current increases roughly exponentially, and secondly because electrons having lost a large amount of energy see a much higher barrier than "elastic electrons". To overcome this difficulty we use a logarithmic derivative $\frac{1}{\sigma} \frac{d\sigma}{dV}$ (where $\sigma = \frac{dI}{dV}$) a third derivative $\frac{d}{dV} (\frac{1}{\sigma} \frac{d\sigma}{dV})$.
- Finally, under high bias, junctions become particularly noisy and unstable. We shall describe an averaging method, using a multichannel analyser, that is far more efficient than a lock-in detector alone in reducing noise problems.

Two types of systems were studied:

- <u>Rare earth oxides</u> have very low lying electronic levels (below 1 eV), but the transitions are strongly forbidden optically (oscillator strengths of $\simeq 10^{-6}$) for parity conservation reasons. A. Adane observed these transitions in Er - Er Ox - Pb and Ho - HoOx - Pb junctions, with conductance variations of 10^{-3}, that is stronger than predicted by analogy with optical spectroscopy.

- <u>Some large molecules</u>, such as dyes also have low lying transitions. We chose a certain number of molecules with either optically allowed singlet-singlet or optically forbidden singlet-triplet transitions in our energy range. We were able to observe both types of transition, with roughly the same conductance variations of a few 10^{-3}. For example we situate the S_0-T_1 transition of carotene at about 1.3 eV. To our knowledge it had not been observed optically.

Our main conclusion is therefore that an "optical model" such as that of SCALAPINO and MARCUS is insufficient to explain electronic transitions. We believe that an analogy with the excitation of molecules by low energy electron impact is more adapted, exchange interaction allowing the single-triplet excitations to take place.

We also find that our peaks are wider than those of the optical spectrum of the dilute molecules. But in several cases we found the same width in the optical spectrum of the molecules in a thin film. The widening would seem to be due to the disordered environment of the molecule, and to not be directly associated with the method.

In conclusion, we have shown that IETS allows the observation of electronic transitions, with optical selection rules lifted. The width of the peaks is a problem, but it might be resolved with an improved method. A proper theory remains to be established, but it is probably analogous to that of low energy electron impact.

IETS has proven to be highly useful in the study of phonons and vibrational transitions [1,2,3], and it seemed worthwhile extending the method to the observation of electronic transitions for several reasons. First, one can expect the electronic energy levels of molecules to be particularly sensitive to their binding to the oxide layer of tunnel junctions, and to give useful information as adsorption and catalysis problems. What's more, reasoning by analogy with optical spectroscopy electronic transitions with oscillator strengths up to .1-1 (against $\simeq 10^{-5}$ for vibrational transitions) could be expected to give enormous conductance variations.

This is very promising, but in practice a certain number of difficulties arise:

The main one is due to the fact that our tunnel junctions break down under biases over 2-3 V, whereas most molecular electronic transitions are way above the corresponding energy. Having tried unsuccessfully to raise this limit with high barrier junctions (such as Li-LiF-Pb which were always short circuits, and Al-Ox-Au which couldn't be doped properly) we had to restrict ourselves to Al-Ox-Pb and Mg-Ox-Pb junctions, and choose systems with electronic levels below 2eV: rare earth oxides and large molecules.

Even within that limit, the inelastic signal is weakened by two causes:

a) the junction current-voltage characteristic ceases to be linear and becomes roughly exponential beyond .5 eV. Thus the elastic background will tend to drown the small increase in current due to inelastic phenomena.

b) when an electron loses a certain amount of energy, it meets a barrier that much higher. As we shall see this strongly attenuates the inelastic current.

For both these reasons, the inelastic current is a particularly small fraction of the total current.

Finally, under high bias, the junctions become extremely noisy and unstable (Fig.2) and the spectra are full of false peaks.

But, as we shall see, we were able to overcome these difficulties.

1. Experimental

The two points of interest are:

a) the derivations [4]

We use a logarithmic derivative $1/\sigma \, d\sigma/dV$ (where $\sigma = dI/dV$. One can show (Fig.1) that this function contrary to d^2V/dI^2 doesn't attenuate inelastic peaks at high bias. A second modulation at a low enough frequency (20 Hz) to pass the first lock-in detector filters, gives us a third derivative $d/dV \, (1/\sigma \, d\sigma/dV)$.

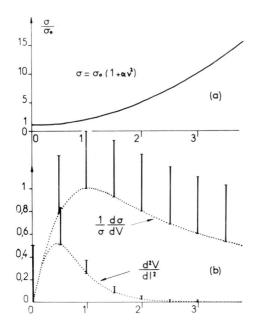

Fig.1 A conductance variation of fixed value is attenuated at high energy by d^2V/dI^2, but not by $1/\sigma \, d\sigma/dV$

b) the averaging [5]

We were troubled by very large amplitude, low frequency noise, to the junction that produced "false" peaks on the spectra (Fig.2).

The lock-in detector is quite capable of removing this type of $1/f$ noise, so we use a multi-channel analyser to add a large number of rapidly swept curves drastically improving the signal/noise ratio (Fig.3).

This averaging system proved invaluable in observing electronic transitions.

2. Rare Earth Oxides

As we said, we are restricted to systems with electronic transitions below about 2 eV. One possibility, examined by ADANE et al. [6], was to use rare earth oxides which have 4f-4f electronic [7] transitions below 1 eV. Unluckily, they are optically forbidden (oscillator strengths of $\simeq 10^{-6}$) for reasons of parity conservation.

The oxides formed the insulating layer of the junctions, for example $Er - Er_2O_3 - Pb$ or $Ho - Ho_2O_3 - Pb$. The transitions were observed (Fig.4)

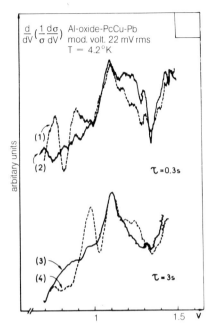

Fig.2 Low frequency, high amplitude noise is not eliminated by increasing the time constant of the lock-in detector

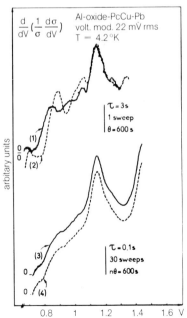

Fig.3 Taking the sum of a large number of sweeps (lower curves) virtually eliminates noise. The total time of measurement is kept constant

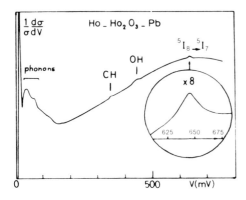

Fig.4 Tunnel spectrum of the
$^5I_8 \to \, ^5I_7$ transition in Ho_2O_3

producing a conductance variation of $\simeq 10^{-3}$, which is at least an order of
magnitude greater than obtained by putting the oscillator strength of 10^{-6}
into SCALAPINO and MARCUS' dipolar model [8]. A different theory for the
excitation of <u>electronic</u> transitions is of course necessary as the other
series of experiments will confirm.

3. Large Molecules

A certain number of complex molecules have electronic levels below 2 eV. We
chose the following molecules, with both optically allowed singlet-singlet
transitions (oscillator strengths .1 to 1) and optically forbidden singlet-
triplet transitions (oscillator strength $\simeq 10^{-3}$)

  - Xenocyanine S_0-S_1 at 1.3 eV   (Fig.5)
  - Tetracyanine S_0-S_1 at 1.3 eV
  - Bis (-4 - Dimethylaminodithiobenzil)-nickel
             S_0-S_1 at 1.25 eV   (Fig.6)
  - Pentacene S_0-T_1 at 8 eV and S_0-S_1 between 1.9 and 2.1 eV
  - Copper Phthalocyanine : S_0-T_1 at 1.15 eV
                  and S_0-S_1 at 1.8 eV (Fig.7)
  - Carotene S_0-T_1 not yet observed optically but expected to be
             below 1.5 eV (Fig.8)

Our main finding [9,10] is that optically forbidden and optically allowed
transitions can be seen with roughly the same intensity (the conductance
variation is of a few 10^{-2}). This lifting of optical selection rules is
particularly useful: for example, we saw the S_0-T_1 transition of carotene
at about 1.3 eV [11]. To our knowledge, it had not been observed optically.

We also find that our peaks are much wider than those an optical spectra
in dilute solution.

But for several molecules, we found that the peaks in the optical spectrum
in a thin film about as wide as our tunnel peaks [12] (see Fig.5). This seems
to show that the widening is due to the disordered environment of the molecules
and not to the method itself. It would be useful to investigate the matter
further, specially as the width restricts the applications of the method.

We also note that the peaks are small: the conductance variation is only
10 times that observed for vibrational transitions whereas the oscillator

206

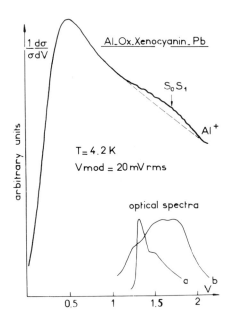

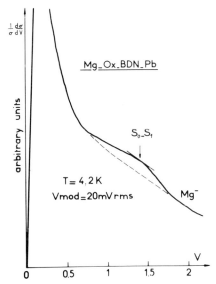

Fig.5 Tunnel spectrum of xenocyanine. The optical spectrum is shown: a) in dilute solution and b) in thin film

Fig.6 Tunnel spectrum of BDN

strengths are in a ratio of ≃ 10 000. This can be explained by the attenuation of the inelastic current mentioned above in point b.

A rough calculation showed us that, of two transitions of equal intensity (same matrix element) but one at .4 eV and the other at 1.4 eV, the higher

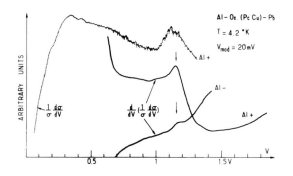

Fig.7 Tunnel spectra of the S_0-T_1 transition in Copper Phthalocyanine. A third derivative was necessary to see the transition in both polarities

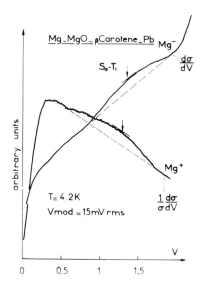

Fig.8 Tunnel spectra of the S_0-T_1 transition in β -carotene

was weakened by a factor $\simeq$ 30. Taking into account the number of vibrational transitions per molecule we find that electronic transitions appear roughly 6×10^3 times stronger than vibrational ones, that is approximately with the strength predicted for an optically allowed transition.

4. The Need of a Theory

From what we have seen it appears that a model such as that of SCALAPINO and MARCUS [8], or even that of KIRTLEY, SCALAPINO and HANSMA [13] very useful for interpreting vibrational spectra, is not adapted to electronic transitions. We feel that the situation is analogous to that of low energy electron impact, where at wide angles, singlet-triplet transitions are excited [14]. MATSUZAWA [15] for example has shown on benzene that an exchange interaction between the incident electron and an electron of the molecule, allows singlet-triplet transitions to take place. The calculation will be more complicated, but probably not fundamentally different when the incident electron is tunneling.

208

5. Conclusion

Electronic transitions are particularly interesting to study by I.E.T.S., because of their sensitivity to adsorption conditions, and because optical selection rules are lifted. A proper theory remains to be established but we believe it will be close to that of low-energy electron impact. For the time being, the width of the peaks observed is a handicap, but one can hope that some device can be found to avoid this. If this is done, I.E.T.S. of electronic transitions should bring a good contribution to the understanding of adsorption and catalysis problems.

[1]Université Paris VII - Tour 23 - 2, Place Jussieu - 75221 Paris Cedex 05

References

1 J. Lambe, R.C. Jaklevic: Phys. Rev. 165, 821 (1968)
2 J. Klein, A. Léger, M. Belin, D. Défourneau, M.J.L. Sangster: Phys. Rev. B 7, 2336 (1973)
3 P.K. Hansma: Proceedings of the 14th conference on Low Temperature Physics, Helsinki (1975)
4 J. Klein, A. Léger, B. Delmas, S. de Cheveigné: Rev. Phys. Appl. 11, 319 (1976)
5 A. Léger, B. Delmas, J. Klein, S. de Cheveigné: Rev. Phys. Appl. 11, 307 (1976)
6 A. Adane, A. Fauconnet, J. Klein, A. Léger, M. Belin, D. Défourneau: Sol. St. Comm. 16, 1071 (1975)
7 N.T. McDevitt, A.D. Davison: J. Opt. Soc. Am. 56, 636 (1966)
8 D.J. Scalapino, S.M. Marcus: Phys. Rev. Lett. 18, 459 (1967)
9 A. Léger, J. Klein, M. Belin, D. Défourneau: Sol. St. Comm. 11, 1331 (1972)
10 S. de Cheveigné, J. Klein, A. Léger, M. Belin, D. Défourneau: Phys. Rev. B 15, 750 (1977)
11 P. Mathis: Ph.D. thesis, University or Orsay (1970)
12 P. Devaux, M. Schott, M. Lazerges: Phys. Stat. Sol. 4, 43 (1964)
13 J. Kirtley, D.J. Scalapino, P.K. Hansma: Phys. Rev. B 14, 3177 (1976)
14 J.P. Doering, A.J. Williams III: J. Chem. Phys. 47, 4180 (1967)
15 M. Matsuzawa: J. Chem. Phys. 51, 4705 (1968)

Light Emission From Inelastic Tunneling – LEIT

John Lambe and S.L. McCarthy

Research Staff, Ford Motor Company
Dearborn, MI 48121, USA

ABSTRACT

It has been shown that inelastic tunneling excitation of surface plasmon
modes can result in light emission when electrodes are properly prepared.
The light emission establishes a fundamental quantum relationship between
the maximum optic frequency and the applied voltage. Such a process can be
used as a basis for spectroscopy when modulation-derivative techniques are
used. The derivative method is very similar to that used in IETS except it
is a second derivative of a photo detection current that is measured. We
term this LEIT spectroscopy (light emission by inelastic tunneling). We
will discuss the physical processes underlying the LEIT effect as well as
possible applications. Some of the key factors of this light source are:
1) the precision of the relationship between voltage and maximum optical
frequency; 2) mechanics for coupling out light for excited surface plasmon
modes; 3) the optical range over which such sources can be operated.

Based on the theory evolved by IETS, one would expect that the precision
of the quantum relationship between voltage and light should be temperature
dependent. We have examined this effect between 200°K and 4.2°K and find
the expected broadening of second derivative spectra with temperature. The
broadening is consistent with the 5.4 KT expected from simple theoretical
considerations. Some problems arise from voltage distribution across the
junction itself since this artifically broadens the second derivative. We
believe that this problem can be solved by using a superconducting grid
over the outer electrode.

It was clear in the early phases of this work that techniques such as
electrode roughening were required to outcouple light from surface plasmon
modes. We have recently demonstrated other methods for inducing outcoupling.
For much of the work currently being done, for example, we have used silver
as the outer electrode and have developed etching processes for producing
outcoupling. These techniques are now fairly reliable and produce external
quantum efficiency between 10^{-5} and (at best) 10^{-4} photons per electron.
These devices have been operated over the spectra range of 6eV (200NM) to
1.2 eV (1000NM, in the near infrared). The U.V. limit is due to voltage
breakdown problems as well as surface mode limitations. We have not examined
the emission at lower energy than 1.2 eV due to limitations of our experimental apparatus. The emission should extend well into the infrared.

It has been shown [1,2] that inelastic tunneling excitation of surface plasmon modes can result in light emission when electrodes are properly prepared.
The light emission establishes a fundamental quantum relationship between

maximum optical frequency and applied voltage. Such a process can be used as a basis for spectroscopy when modulation-derivative techniques are used. The derivative method is very similar to that used in IETS, except that it is the second derivative of a photodetector current that is measured. We have termed this LEIT spectroscopy (Light Emission by Inelastic Tunneling). We will discuss the physical processes underlying the LEIT effect as well as possible applications. Some of the key features of this light source are (1) the precision of the relationship between voltage and maximum optical frequency, (2) mechanisms for coupling out light from excited surface plasmon modes, (3) the optical range over which such sources can be operated.

The basis of this light source is the inelastic excitation of the electromagnetic modes of the tunneling junction. Thus, instead of exciting molecular vibrations we will excite electronic "vibrations". These are very curious modes. The evidence for their presence comes primarily from theory [3], but TSUI [4] had done an IETS experiment some years ago on semi-conductors which indicated such modes were there. One also knows from work on Josephson A.C. effect that "slow" waves exist in superconducting tunnel junctions. This slow wave also can exist in normal metal junctions [3].

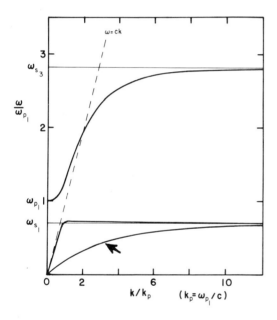

Fig.1 Dispersion curves for a metal-insulator-metal tunneling junction. (Aℓ-Aℓ$_2$O$_3$-Ag for example). Arrow indicates "slow wave" branch of interest in present context

The dispersion curve for such modes is shown in Fig. 1. One notes a continuum up to an asymptotic limit at the surface plasmon frequency. These modes are well beyond the light line so they are manifestly "nonradiative".

Our approach was to produce microstructure in the tunneling junction that could scatter such modes and thereby make them radiative. Thus, one hopes to excite such modes with tunneling electrons and see light.

The production of an appropriate microstructure is necessary in order to observe light. We have developed a number of methods to do this. In Al-Ag junctions one can expose the junction to halogens and then heat for a minute or so in air at 150°C. The Ag electrode becomes rough and porous. Microstructures can also be produced on top of the tunneling junction [2]. The idea is, of course, to couple the "trapped light" being generated in the junction to the free space photon field.

When a junction is properly prepared, one can see light being emitted from the junction when voltage is applied. But now there is a very striking effect which confirms the quantum mechanical origins of this light. We cannot excite an excitation unless the voltage source provides the appropriate

voltage. That is to say, there has to be a fundamental quantum limit on the range of excitations. This meant that as you begin to turn up the voltage, the junction should first look red as we excite all the modes all the way up to two volts, for example. Then, as we go further, it should look orange, then blue and at last a nice, uniform white. That is exactly how the experiment looks. That, of course, is the delight of the effect. One has to be careful and do many other tests of this kind.

Optical spectra are taken with a conventional grating spectrometer. This shows light output versus the wavelength, which displays the spectral character of this light source at a fixed voltage bias. Of course we must see some quantum effect and this is shown in the following way. Figure 2 shows the spectral output for junctions at various fixed applied voltages. One notes that if 2.6 volts is applied, the spectrum stops at a frequency ν corresponding to $h\nu$ equal to 2.6 eV. We apply 3 volts and the light stops at 3 eV. Similarly, for 4.06 volts, it is very clear that the anticipated quantum effect has been observed. With Al-Au junctions we have observed this quantum cutoff into the UV at 6 volts (about 2000 Å).

Thus, the LEIT source seems to be a new way to produce light, and it has its own special quantum relationship. It is bright enough to be viewed in a dimly lit room. There are many technological issues such as efficiency (we have achieved quantum efficiency to 10^{-4} photons per electron) and durability. We will not discuss these now, but rather focus our attention on the physics of this device.

Since we are used to doing spectroscopy, we thought we would take some of the ideas of the technology stimulated by the one subject and use it in this new subject of light generation. The first thing we note is that there is a whole continuum of excitations and so we can't detect them by IETS. But now that we are getting them to emit, we can select out of that continuum any narrow range of excitations we want to look at. Now we can get into the tun-

212

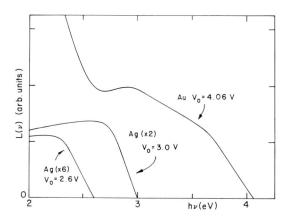

Fig.2 A plot of $L(\nu)$, the photon flux per unit frequency bandwidth, versus $h\nu$ for $A\ell$-$A\ell_2O_3$-Ag and $A\ell$-$A\ell_2O_3$-Au junctions at different applied voltages. Junctions were at 77°K

neling spectroscopy business. We want to show how we can use these ideas to test the physics of this device. It is very important when one is trying to do something in a technological sense to be as careful as one can to understand the underlying physics. What we want to do then is to show that the LEIT source behaves in a manner very much as we would have expected for a small group of excitations. What we will do then is to use an optical filter, say at some fixed frequency, and examine the light that passes out through the filter from the light source as we change the voltage. Now we can go through all the rules of inelastic tunneling and see how things behave. What's rather nice is for the first time we can see an inelastic tunneling channel opening up without any derivatives. You see it very directly in a photomultiplier current. Figure 3 shows what happens when we look at a tunneling junction that has been made to be an emitter. We look at it through a narrow band filter, centered at 6,000 Angstroms which is 2.006 eV. We see that at low voltage we are not generating any excitations that can get through that filter. As we apply the voltage at which this filter will start passing light, the photo response of the detector starts up and now we know we are generating excitations at this voltage. One can compare the optical filter with the threshold voltage reading and it comes out very close. We can apply the voltage with either polarity and the same threshold effect is seen. This accords with what we would expect from what is known about IETS. If it's a genuine tunneling process, we are talking about wave functions overlapping so polarity shouldn't matter as this excitation sort of permeates the barrier. At low voltages one will expect to see the excitations that permeate the entire barrier. (That's not true in the higher voltage region. Excitations tend to glue one onto the silver.) Now we can select a different filter, for example, a 5,200 Angstrom filter which should turn on at 2.304 eV, and, of course, that happens as shown in Fig. 4.

As soon as we saw this, we wanted to know how sharp this quantum effect was. The relation $h\nu = eV$ is very fundamental and so one would like to know

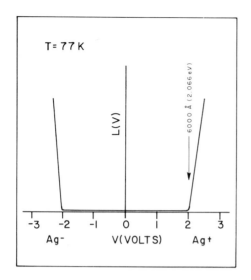

Fig.3 Light intensity for light generated at a frequency corresponding to 2.066 eV as a function of the applied voltage. This shows the "turning" on of an inelastic channel. Note turn on voltage is given by hν = eV. An optical filter was used to select frequency

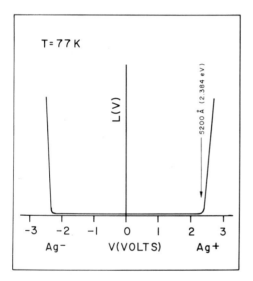

Fig.4 Light intensity for light generated at a frequency corresponding to 2.384 eV

how precisely the threshold can be determined. That, of course, comes out directly from all the discussions that have gone on about the precision or resolution of ordinary IETS. And I'll just refer to an old picture [5] (Fig. 5). It shows the essential feature we are looking for. What we are talking about is how sharply does an inelastic channel turn on? What this curve tells us is that if we could select an infinitely sharp set of inelastic channels, we know that the turn on would be broadened out slightly. It's advisable to put such questions in terms of second derivative. So one would

214

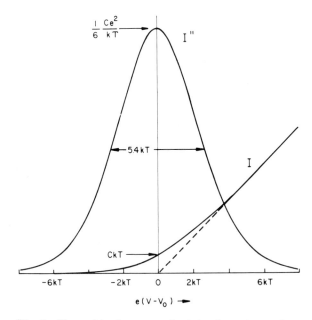

$$\frac{1}{6}\frac{Ce^2}{kT}$$

I "

5.4 kT

I

CkT

-6kT   -2kT   0   2kT   6kT

$e(V-V_0) \longrightarrow$

Fig.5 Theoretical curves for the "turning on" of an inelastic channel and the second derivative [5]. Second derivative half-width is 5.4 kT for an infinitely sharp excitation

say we should see a line width, a resolution width, of 5.4 kT just as for regular IETS.

In principle an extremely sharp optical filter would be used to select a very narrow distribution of excitations. One can, however, fold in the finite filter width by noting that the second derivative width will be given by

$$\Delta = \sqrt{(\Delta_F)^2 + (5.4\ kT)^2}$$

where Δ_F is the half width of the optical filter. Second derivative measurements were carried out at various temperatures between 200°K and 4.2°K. The expected thermal broadening effect is seen just as one would expect.

As an example of the second derivative spectral method, Fig. 6 shows the LEIT spectra of an optical filter centered at 2.4 eV. We also show the optical transmission of the filter. The second derivative of the photodetector current is essentially the same as the optical spectra. This was done with the LEIT device at 77°K. The filter half-width was 53 meV. The LEIT second derivative half-width is 66 mV. Theory predicts 64 mV, so the agreement is good.

In Fig. 7 we show a LEIT spectra taken at 4.2°K. The optical filter had a width of 9 meV and we see a width of 12 mV. Thus there is still some extraneous broadening. We believe that the finite sheet resistance of the Ag

215

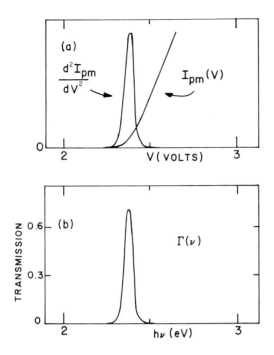

Fig.6 (a) Photomultiplier current I_{pm} and its second derivative with respect to voltage plotted versus voltage for an Ag junction at 77°K. (b) The transmission function for the filter used to view the junction

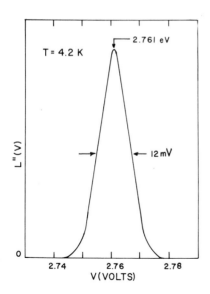

Fig.7 Second derivative of light intensity through a fairly narrow optical filter. The junction was at 4.2°K

electrode is responsible for this. Nonetheless the second derivative spectra is quite sharp and shows the filter position very well. At 77°K the line was broadened out to about 35 mV as one would expect. Thus the second derivative technique in LEIT spectra has the familiar temperature-dependent resolution as IETS. This may be unfortunate from the point of view of useful spectroscopy, but it is nice to see that the physical principles are as we expected.

In order to give some perspective on the LEIT spectra, in Fig. 8 we show the data of Fig. 7 on a scale which encompasses the complete spectral range.

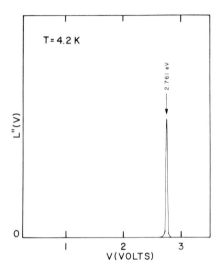

Fig.8 Data of Fig. 7 plotted on a scale which shows the total spectral range

One then gets some feeling for the potential resolution of LEIT spectroscopy.

In closing, I wish to thank Professor Wolfram and the people here at the University of Missouri on behalf of all the participants at this Conference. We all are grateful for their kind hospitality and their efforts to make this meeting a very interesting and successful one.

References

1 John Lambe, S.L. McCarthy: Phys. Rev. Lett. 37, 923 (1976)
2 S.L. McCarthy, John Lambe: Appl. Phys. Lett. 30, 427 (1977)
3 E.N. Economou: Phys. Rev. 182, 539 (1969)
4 D.C. Tsui: Phys. Rev. Lett. 22, 293 (1969)
5 John Lambe, R.C. Jaklevic: Phys. Rev. 165, 821 (1968)

VII. Elastic Tunneling

Zero Bias Anomalies in Tunneling: A Review

E.L. Wolf

Ames Laboratory-ERDA and Department of Physics, Iowa State University
Ames, IA 50011, USA

ABSTRACT

The measured conductance $\sigma(V)$ of a tunnel junction between two metals may depart from the simple idealization of a constant σ_0, independent of applied bias voltage V for several more or less basic reasons, which will be the topic of this review. First, a nearly temperature independent quadratic increase in conductance $\sigma = \sigma_0 + \sigma_1 (V-V_0)^2$ is often observed, and may reasonably be ascribed to the distortion of a rectangular or trapezoidal insulating tunnel barrier with variation of V. An offset V_0 of the structure may arise from a built-in electric field across the barrier. Such an effect is almost an intrinsic feature, therefore, of a tunnel junction. Two other frequently observed temperature-dependent structures which occur precisely at V = 0, on the other hand, are evidence for tunneling mechanisms additional to direct elastic tunneling and are associated with additional localized electron states in or near the barrier. A "giant resistance peak," or conductance minimum has convincingly been ascribed to two-step tunneling across the barrier via real intermediate states occurring on small metal particles imbedded in the barrier. Variations of this effect have been seen in other contexts, including that of metal-semiconductor tunneling. The narrower and weaker conductance peak can be attributed to magnetic scattering of tunneling electrons from paramagnetic impurities at the edge of the barrier, effects which may be described as Kondo and spin-flip scattering across the tunneling barrier. Finally, the effect of a narrow conduction band in leading to an asymmetric tunnel conductance will be briefly described.

For the research worker interested in inelastic electron tunneling spectrocopy, the ideal background behavior of an undoped metal-insulator-metal (MIM) tunnel junction would be a perfectly Ohmic response, described by dI/dV = σ = constant and $d^2I/dV^2 = 0$. While such desirable baseline behavior cannot strictly be attained, it can be approached, and it may be helpful to briefly review what is known about several types of departure of the undoped MIM tunnel spectrum from the idealized baseline $d^2I/dV^2 = d^2V/dI^2 = 0$. Identification of these characteristic effects may provide helpful clues in deducing the kind of barrier defects which result from a given fabrication process.

First, the strongest "zero-bias anomaly" that one is likely ever to observe is the superconducting energy gap, e.g., of a Pb counter-electrode at 4.2°K: this structure in dI/dV is symmetric about V = 0, including the gap region itself where $(dI/dV)_s/(dI/dV)_n \lesssim 10^{-3}$ for $|eV| < \Delta$ = 1.3 meV; the strong peaks in dI/dV at eV = $\pm \Delta$; and the phonon range of about 10% deviations in dI/dV extending (in Pb) to about ±10 meV, with harmonic structure to 20 meV or more. Observation of this structure, including particularly the ratio $\sigma_s/\sigma_n < 10^{-3}$ for eV< Δ, is proof of a good tunnel junction; and the IETS energy resolution is improved by the presence of the sharp peaks in the superconducting density of states at $\pm \Delta$. On the other hand, the large

Pb phonon structures may complicate the d^2V/dI^2 spectra if one is interested in IETS peaks below 50 mV. In this case, following P. K. HANSMA [2], one might wish to use a bridge arrangement containing <u>two</u> Pb-I-M junctions (only one being doped) to balance out the undesired Pb phonon structure.

Another inherent deviation from the idealization $d^2V/dI^2 = 0$ comes from the fact that changing the bias voltage across the junction must change the "shape" of the potential barrier $\phi(x,V)$ through which tunneling occurs, and thus change the transmission probability. This situation is sketched in Fig. 1, taking a trapezoidal barrier model

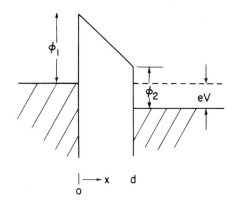

Fig.1 Sketch of the trapezoidal barrier model under applied bias voltage

$$\phi(x,V) = \phi_1 + \frac{x}{d}(\phi_2 - eV - \phi_1) \tag{1}$$

where the barrier heights $\phi_{1,2}$ are related to the metalic work functions, d is the barrier thickness and V the applied bias voltage. Calculation of the tunneling current I (for unit area), using the WKB approximation, gives

$$I = \frac{2e}{h} \sum_{k_y,k_z} \int_{-\infty}^{\infty} \exp[-2\int_0^d \mathcal{K}(x,E_x,V)dx] \cdot \tag{2}$$
$$\cdot [f(E)-f(E-eV)]dE_x$$

where

$$\mathcal{K}(x,E_x,V) = \{\frac{2m}{\hbar^2}[\phi(x,V)-E_x]\}^{1/2} \quad . \tag{3}$$

Here x is the tunneling direction, and E_x the kinetic energy associated with the x-motion; the f's are Fermi functions to require that tunneling occurs only from occupied to unoccupied states. A numerical study of the conductance dI/dV resulting from this model indicates [3] an approximately parabolic increase in conductance from a minimum which may be offset from V = 0; e.g., by 70 mV for asymmetric barrier heights $\phi_1 = 3$ eV, $\phi_2 = 1$ eV. The increase in conductance from this effect (change of the barrier shape with bias) may amount, with reasonable parameters, to doubling dI/dV (relative to the dI/dV minimum value) in a 300-600 mV change of bias. The calculated features are relatively insensitive to inclusion of the image force or to use of an exact transmission factor instead of the WKB factor [3]. The authors concluded that a reasonable description of the background conductance for typical undoped barriers could thus be obtained, but that dI/dV background shapes corresponding to heavily doped junctions (which may have offsets of the dI/dV minimum by ~250 mV) are not so well predicted. An appropriate extension of the barrier model to include adsorbed molecules has recently been reported by WALMSLEY et al. [4], who obtain excellent model fits to experimental measurements on doped junctions.

There is one other inherent kind of departure from the idealized limit $d^2V/dI^2 = 0$, which is worth mention. This is a very weak ($\Delta\sigma/\sigma \sim 10^{-3}$) and narrow (~1 mV) conductance minimum which occurs below 10 K in ideal (normal metal) MIM junctions with thin and highly transmitting oxide barriers, simply because the electron distribution at the Fermi surface of a very pure electrode may be driven slightly out of thermal equilibrium by the tunnel injection process [5]. Neglecting impurity scattering, the "hot" injected electron at E = eV must depend upon phonon emission to leave the particular $\vec{k}$ state into which it is injected and finally to thermalize, and relatively few phonon modes (and unoccupied electron final states) are available at small (millivolt) energies, leading to long electron life times as E → 0. TROFIMENKOFF et al. [5] show, by straightforward consideration of electron relaxation rates, that the resulting conductance minimum has the form (for T = 0)

$$\frac{dI}{dV}(V) \propto D \left[1 - \frac{D/\tau_0}{\frac{1}{\tau_i} + \frac{\pi}{\hbar} \alpha^2 (eV)^2} \right] . \tag{4}$$

In the bracket term, D/τ_0 represents the injection rate of electrons across the barrier, while the two denominator terms are, respectively, relaxation rates by impurity scattering (τ_i^{-1}) and by emission of phonons of all energies up to $\hbar\omega = eV$. Favorable comparison of this theory, including the temperature dependence (omitted here) with experiment is obtained [5]. For Al-I-Al junctions of 2000Å electrode thickness, D/τ_0 is found to be 7×10^6 sec^{-1} and τ_i is given as 3.5×10^{-10} sec.

Finally, we will describe two types of zero bias anomaly which may arise from interaction of tunneling electrons with point defects in the barrier or at the edge of the electrode, and which are thus indications that the junction is imperfect. The two behaviors are a temperature- and magnetic field-dependent conductance peak of 1% to 10% strength and typically millivolt width, associated with magnetic moments in the barrier; and a frequently large ($\Delta R/R \sim 1$) and strongly temperature-dependent, but magnetic field-independent zero-bias resistance peak, associated with real, localized intermediate states inside the barrier, a process which is sometimes referred to as two-step tunneling [6]. These two types of anomalies may well arise in attempts to dope barriers with small metallic particles (especially of ferromagnetic metals), as in the recent studies [7] of chemisorption of CO on alumina-supported Ni and Rh particles. Thus, a discussion of the possible associated tunneling anomalies seems timely.

The general situation in the case of the magnetic conductance peak is indicated in Fig. 2, in which the tunneling electron wave function overlaps a localized wave function with which is associated an unpaired spin $\vec{S}$ and associated magnetic moment $\vec{\mu}$, whose magnitude is $g\mu_B$ with μ_B the Bohr magneton, 9.27×10^{-21} ergs/oersted; the g-factor is 2.00 for a free electron. Of course, if a magnetic field $\vec{H}$ is present the magnetic moment can take one of (2S+1) quantized orientations split in energy by the Zeeman energy $g\mu_B H$. Now the overlap of the wave functions means that an exchange interaction

222

$$\mathcal{H}^* = -2J\ \vec{S}\cdot\vec{\sigma} \tag{5}$$

exists between the local moment $\vec{S}$ and the spin $\vec{\sigma}$ of the tunneling electron, and it is this magnetic interaction that leads to the zero bias anomaly. The simplest interaction process taking place during tunneling is the mutual spin-flip scattering process in which the spin of the localized moment is reversed, a process which for $H \neq 0$ will require an energy $g\mu_B H$, and thus is associated with a bias threshold $|V| \gtrless g\mu_B H/e$. This is another example of IETS, and the tunneling technique has indeed been used to measure the g-value of localized moments [8]. The zero-field conductance peak, however, is a more subtle effect, based on the same magnetic interaction, but involving only virtual spin flips: it is mathematically similar to the scattering process described in higher order perturbation theory by KONDO in 1964 [9]. This scattering process is characterized by a $-\ell n\ T$ temperature variation at $V = 0$ (at least for T above the KONDO temperature, T_K). In fact, the voltage dependence of dI/dV (once the background conductance is subtracted) is also logarithmic. This effect has probably been most thoroughly understood as it occurs in metal-semiconductor Schottky-barrier tunnel junctions [8], in which case the magnetic moments are inherently present at the inner edge of the depletion region and correspond to localized neutral hydrogen-like donor impurities. The (H = 0) data for dI/dV in Fig. 3 was obtained from such metal-semiconductor junctions [8]; the solid curves are an interpolation function

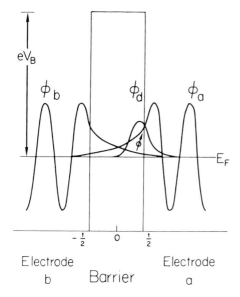

Fig.2 The situation which leads to the magnetic scattering conductance peak. The localized wave function is associated with a magnetic moment $\vec{\mu}$

$\propto \ln[(eV)^2 + W^2]$ which is suggested by the perturbation theory results [9]. It is important to understand here that the tunneling electron does not localize upon the magnetic moment; in fact, in the H=0 case even the spin-flip of the local moment is only a virtual process (i.e., one that persists only for a time $\leq \hbar/\Delta E$). These effects have also been observed in junctions doped with submonolayer amounts of magnetic metals such as Fe, Ni, Co, etc.

In those cases where localization of the tunneling electron at some intermediate position in the barrier does take place, this can produce a large

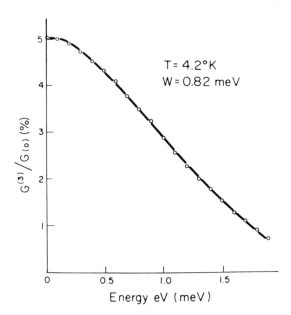

Fig.3 The conductance anomaly (points) as a function of bias, compared with the perturbation theory (solid curve). The value at V=0 varies with temperature as -ln T. Magnetic moments in a Si Schottky barrier after WOLF and LOSEE [8]

temperature-dependent resistance peak at V=0. Two examples which we mention of this effect are the deliberately introduced small (100Å-1000Å) metal particles studied by ZELLER and GIAEVER [10], and a case involving microscopic localized states in a compensated semiconductor [11]. The latter case is of some interest because the varible-range assisted tunneling behavior of MOTT [12] was observed.

The ZELLER-GIAEVER [10] geometry forces the average electron in tunneling from one Al electrode to the other to do so in two steps (see Fig.4): from

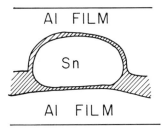

Fig.4 The geometry of an imbedded small Sn particle in the tunneling barrier, after ZELLER and GIAEVER [10]

the bottom through Al_2O_3 into the Sn particle; and then from the Sn particle through the upper Sn oxide into the upper Al film. In their classical analysis of this situation, ZELLER and GIAEVER [10] estimated, from electron microscope pictures, the sizes and then the capacitances C of the particles - that is, the charge that can be stored on the particle per volt of electric potential V. From this was determined the energy necessary to charge the particle by <u>one</u> electron, which would normally be $\Delta E = e^2/2C$ (typically a few milleV), but in this case is

$$\Delta E = \frac{1}{2}(\frac{e}{C} + V_D)^2 - \frac{1}{2} V_D^2 C \tag{6}$$

because the particle in its uncharged state may still have a residual voltage V_D, assumed to be distributed equally in the range $- e/2C < V_D < e/2C$. This then gives a uniform distribution of threshold energies $|eV| < e/C$ allowing tunneling across the junction through the Sn particles, which leads to a V-shaped conductance minimum at V=0 such that $\sigma(0) \propto T$, which is the observed temperature dependence. Viewed in the resistance, dV/dI, the effect is a dramatic peak as is shown in Fig. 5, after ZELLER and GIAEVER [10]. The basic idea is simply that at V=0 the activation energy to localize on a particle is not available, while by V=e/C, all particles are energetically

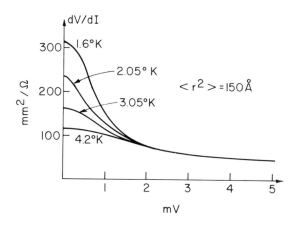

Fig.5 The resistance peak observed by ZELLER and GIAEVER [10] as a function of temperature

accessible and the full conductance is reached.

The picture given above extends in a qualitative fashion to microscopic sites (traps) in the barrier or electrode upon which the tunneling electron may localize. Such states do not exist in the band gap region of a perfect insulator, but various microscopic chemical or structural point defects can act as such trapping centers. One situation in which localization can occur is on vacant donor impurity sites in an N-type semiconductor which is partially compensated with acceptor impurities. If there are N_D donors and N_A acceptors per unit volume, then there will be (at T=0) exactly N_A empty donor sites (the traps) which could be momentarily occupied by an externally injected tunneling electron. The energy required for occupation of a given empty donor site will depend on its neighborhood, and will fluctuate from site to site, reminiscent of the distributed parameter V_D of the metal particles. If we imagine such a region lying at one edge of the tunneling barrier, and consider the rate of tunneling across the barrier into this region at zero applied bias voltage, we would expect, following MOTT [12], that the rate is

$$\omega \propto e^{-2\mathcal{K}R} e^{-\Delta E/k_B T} \qquad , \qquad (7)$$

where $\mathcal{K}$ is the decay constant of the wave function, R the distance to the empty donor site, and ΔE the energy deficit which must be obtained from the phonons to reach a given site. MOTT has shown [12] that the <u>most probable</u> tunneling events will involve large R and small ΔE at low temperature, and larger ΔE but smaller R at high temperature: the optimum rate at a given T is given by the "Mott $T^{1/4}$ law"

$$\left. \begin{array}{l} \omega \propto e^{-B/T^{1/4}} \qquad \text{where} \\[4mm] B = 2.1 \left[\dfrac{3}{k_B N(\mu_F)} \right]^{1/4} . \end{array} \right\} \qquad (8)$$

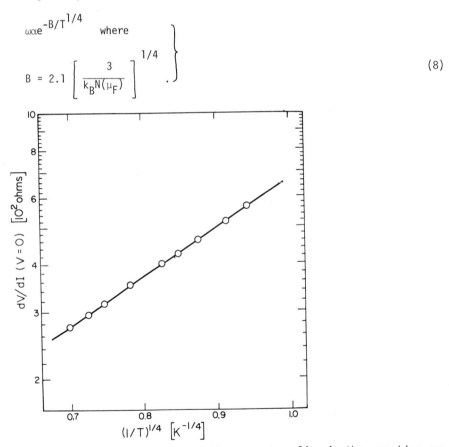

Fig.6 Demonstration of Mott variable range tunneling in the zero bias conductance of a Schottky barrier junction on compensated Si:Sb. The localized intermediate tunneling states occur on the $N_D - N_A$ unoccupied donor sites

Here $N(\mu_F)$ is the density of localized states at the Fermi energy, and k_B is Boltzmann's constant. The predicted temperature dependence then identifies the variable range hopping process which may occur in a tunneling experiment with a volume distribution of localized states in the barrier or at the edge of the electrode. An example of this is given in Fig. 6, corresponding to a Schottky barrier (metal-semiconductor) contact on silicon with Sb donors and Ga compensating acceptors. The localized states occurred at the inner edge of the depletion region in the silicon crystal [11]. As a general rule a conductance minimum with a noticeable temperature dependence below 10K should be suspected as containing localization sites; the particular voltage and temperature dependences will depend upon the details of the situation.

Acknowledgement

This work was supported by the U.S. Energy Research and Developement Administration, Division of Physical Research and Division of Basic Energy Sciences at the Ames Laboratory, Iowa State University.

References

1   W.L. McMillan, J.M. Rowell: in "Superconductivity" R.D. Parks, Editor (Marcel Dekker, Inc., New York, (1969), p. 561
2   P.K. Hansma: paper CA2 in this Conference
3   W.F. Brinkman, R.C. Dynes, J.M. Rowell: J. Appl. Phys. 41, 1915 (1970)
4   D.G. Walmsley, R.B. Floyd, W.E. Timms: Solid State Commun. 22, 497 (1977)
5   P.N. Trofimenkoff, H.J. Kruezer, W.J. Wattamaniuk, J.G. Adler: Phys. Rev. Lett. 29, 597 (1972)
6   A more complete discussion of these topics is given by E.L. Wolf, in Solid State Physics 30, 1 (1975)
7   P.K. Hansma: paper SA1 in this Conference
8   E.L. Wolf, D.L. Losee: Phys. Rev. B 2, 3660 (1970)
9   J. Kondo: Progr. Theoret. Phys. (Kyoto) 32, 37 (1964)
10 H.R. Zeller, I. Giaever: Phys. Rev. 181, 789 (1969)
11 E.L. Wolf, R.H. Wallis, C.J. Adkins: Phys. Rev. B 12, 1603 (1975)
12 Sir Nevill Mott: "Metal-Insulator Transitions" (Taylor and Francis Ltd., London, 1974), p. 34

Superconducting Tunneling

R.C. Dynes

Bell Telephone Laboratories
Murray Hill, NY 07974, USA

ABSTRACT

The general property of electron tunneling into superconducting materials
and the information available from such a technique will be reviewed. The
advantages of using superconducting rather than normal metal electrodes for
IETS will be outlined. Using the Josephson effect, for example, one can
profile the tunneling probability along a junction to determine current distribution. Measuring the superconducting energy gap and the phonon renor-
malized density of excitations the superconductor can serve as extremely
sensitive tests of the quality of the tunnel barriers and if, in fact, tunneling is the dominant conduction mechanism. Artificial barriers have been
often suggested as attractive alternatives to natural oxides. The quality
of these barriers as well as some of the details of the tunneling process
can be implied more easily when the electrodes are superconducting. There
are various other phenomena which are observed when the electrodes are superconducting and what they tell us about the barrier or the electrode will be
reviewed. In cases where these phenomena result in a significant distortion
of IET spectra, techniques to avoid them will be outlined.

Phonon Damping of Virtual Levels in Thick Superconducting Films

W.J. Tomasch

Department of Physics, University of Notre Dame
Notre Dame, IN 46556, USA

ABSTRACT

By virtue of their sensitivity to incoherent scattering, virtual-state resonances observed by tunneling into thick-film elemental superconductors provide a superior probe of the electron-phonon interaction at low energies where relatively few phonon modes participate. In strong-coupling and intermediate-coupling systems, spontaneous phonon emission progressively limits quasiparticle mean-free-paths with increasing energy, causing resonance amplitudes to fall below those predicted for the weak-coupling limit. Accelerated damping of virtual-state oscillations at higher junction biases provides a practical basis for experimental determination of $Im\{Z(\omega)[\omega^2-\Delta^2(\omega)]^{1/2}\}$ which in turn permits $\alpha^2F(\omega)$, $Im\{Z(\omega)\}$ and $Im\{\Delta(\omega)\}$ to be inferred by means of the ELIASHBERG equations. This approach has proven considerably more sensitive than the familiar tunneling method of ROWELL and McMILLAN, and generally tends to augment the latter at low energies. Experimental results obtained with very thick In films exhibit substantial phonon damping compared to those obtained with thick Al films. The coefficinet g_0 appearing in $\alpha^2F(\omega)\dot{=}g_0\omega^2$ determined for In by the present method is found to be substantially smaller than that inferred from earlier tunneling studies, as are the quantities $Im\{Z(\omega)\}$ and $Im\{\Delta(\omega)\}$.

Tunneling With Spin Polarized Electrons

R. Meservey

Francis Bitter National Magnet Laboratory, Massachusetts Institute of
Technology, Cambridge, MA 02139, USA

ABSTRACT

When a superconductor of low atomic number is placed in a magnetic field H
the peaks in the quasiparticle states on each side of the energy gap are
Zeeman split by an amount $2\mu H$ where μ is the magnetic moment of the electron.
In order to see this effect the film must be thinner than about 100 Å and
oriented with its plane in the field direction. Under these conditions the
effect of orbital depairing is small and the effect of spin splitting can
be observed in the tunneling conductance. Al is the most convenient substance in which to observe this Zeeman splitting because its low atomic number leads to little spin orbit scattering and thereby little mixing of the
spin states. In addition, Al films are easily made as thin as 40 Å and
Al_2O_3 forms a good tunnel barrier. The Zeeman splitting of the quasiparticle
peaks allows one to study the properties of the counter-electrode by means
of an electron-spin polarized tunnel current whose polarization is a known
function of the voltage applied to the junction.

The technique has been useful in investigating the spin density of state
of superconductors in high magnetic fields where the effects of electron
paramagnetism can dominate the thermodynamic behavior. Tunneling gives a
direct measure of the amount of spin-orbit scattering in the superconducting
electrodes and therefore their high field properties. Ultra-thin films of
Al, Ga, In, Sn, V, PdH formed by cryogenic deposition have been studied by
this method.

In superconductor-ferromagnetic junctions, the polarization of the tunneling current gives the effective spin density of states of the ferromagnet
at the Fermi energy. Measurements have been made on Fe, Ni, Co and alloys
of Ni with Fe, Cr, Ti, Cu, or Mn. For all these 3d ferromagnets the tunnel
current was found to be positive (predominantly of the majority spin carriers
of the ferromagnet) and approximately proportional to the bulk magnetization.
Measurements have also been made of the rare earth metal Gd, in which the
polarization is also positive.

The capabilities and limitations of this method will be discussed. Possible extensions to high temperatures, higher fields, and inelastic tunneling will be discussed.

My talk mainly concerns <u>elastic</u> tunneling with spin polarized electrons and
describes the history of these experiments over the past six years. The
original purpose was to study the basic physics of high field superconductivity and ferromagnetism. The purpose of my talk in this symposium is to present the results and the techniques with the hope that they may be applicable
to inelastic tunneling. My co-worker in all this work has been PAUL TEDROW
and recently DEMETRIS PARASKEVOPOULOS has joined us. Figure 1 shows the

230

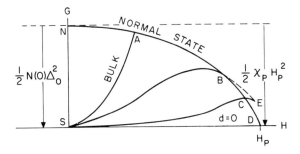

Fig.1 The free energy G of normal and superconducting films as a function of magnetic field H. Curve A for bulk or a thick film increases quadratically with H and has a first order phase transition to the normal state. Thinner films have higher critical fields at B and C. Even a zero thickness film would have a finite critical field H_p because of the quadratic decrease of the normal state G because of its paramagnetism. The phase transition at B is second order, but for thinner films the transition is first order at C or D. E is in a metastable region of supercooling

free energy of a superconducting film as a function of the magnetic field H applied parallel to the film surface. For a bulk superconductor or a thick film (curve A) the free energy increases quadratically with H and its inter-section with the free energy of the normal state defines the critical field H_c. For thinner films (curves B and C) H_c increases and in the limit of zero thickness (line D) the free energy is independent of H. Even so, for a BCS superconductor (by this I mean one in which the spins are paired op-positely in the superconducting state) there is a finite critical field. The normal state free energy decreases with H because of its spin paramag-netism whereas the superconducting state has no spin paramagnetism and its free energy is constant with H, and the intersection with the normal state is at H_p, the paramagnetic critical field. A 40 Å thick Al film approaches H_p closely. H_c being about 50 kG. The transition to the normal state is a first order phase transition. Figure 2 shows the measured critical field of Al films as a function of film thickness d [1]. When d becomes less than about 5000 Å, H_c increases as $d^{-3/2}$ and then flattens out because of para-magnetic limiting.

To make a 40 Å thick film, the Al is deposited on a glass slide held at near liquid nitrogen temperature. The film is then warmed to room tempera-ture and oxidized with a glow-discharge or by a 30 minute exposure to labor-atory air nearly saturated with water vapor. The junction is then completed by depositing a cross counterelectrode. Figure 3 shows the conductance of such a thin $Al-Al_2O_3-Ag$ junction. In zero field one gets the characteristic density of states peaks above the energy gap. In a parallel field of about 10 kG there is Zeeman splitting of the quasi-particle states [2]. States with one spin direction increase in energy and those of the opposite spin direction decrease in energy, and one gets the characteristic splitting of

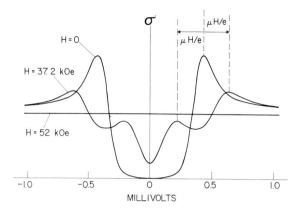

Fig.2 Critical field H_c of an Al film in a parallel magnetic field as a function of film thickness d. For 5000 Å > d > 100 Å, $H \sim d^{-3/2}$. For d < 100 Å, H_c is strongly limited by spin paramagnetism

$2\mu H$ shown in the figure. In the region of the first right hand peak the tunneling electrons are of one spin direction; in the region of the first left hand peak they are of the other spin direction. This would be exactly true if Al were a perfectly spin-paired superconductor. In the real world things are never quite that simple. In Al there is a small amount of spin-

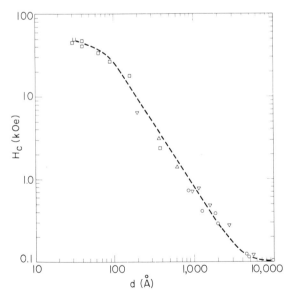

Fig.3 Tunneling conductance of an Al-Al$_2$O$_3$-Ag film as a function of applied voltage. In a parallel magnetic field of 37.2 kOe the quasiparticle density of states is Zeeman split by $2\mu H$ as shown

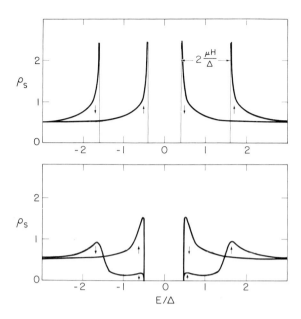

Fig.4 Top curves show theoretical BCS density of states for a paramagnet-ically limited supercondutor in a magnetic field. The lower curves shows the effect on the density of states of a small amount of spin-orbit scatter-ing according to ENGLER and FULDE [3]

orbit scattering so that there is some mixing of the spin states. Figure 4 shows the resulting density of states as calculated by ENGLER and FULDE [3] in the lower part of the figure and the theoretical split BCS density

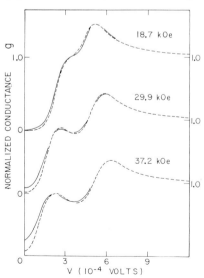

Fig.5 Comparison between experimental tunneling conductance curves and the-ory as calculated by BRUNO [5]

of states for comparison in the upper part. Using the MAKI-FULDE [4] theory
BRUNO and SCHWARTZ [5] have calculated the conductance curves including the
effects of spin-orbit scattering and orbital depairing caused by the applied
field. The result is compared to an experimental curve in Fig. 5. The a-
greement is good and so we can analyze data with some precision.

Using the spin-polarized tunnel currents from superconducting Al we have
investigated the effective spin density of states of ferromagnetic metals
at E_F [6]. If the effective density of states of the ferromagnet are different in the two spin directions we obtain the asymmetrical curves shown in
Fig. 6. Here we show the conductance versus voltage for a Al-Al$_2$O$_3$-Co junction. By analyzing the curves we find a polarization of the tunnel currents

$$P = \frac{j\uparrow - j\downarrow}{j\uparrow + j\downarrow} \approx + 34\%$$

Nickel is about +9%, iron about +44%. The positive sign means that the
tunnel current is predominantly in the majority spin direction in the ferromagnet. In fact it has turned out with these elements and many transition
metal alloys that the sign of P is always positive and the magnitude is
approximately proportional to the saturation magnetic moment of the ferromagnet [7]. This result is shown in Fig. 7.

One experiment which we have done shows the sensitivity of the method.
Figure 8 shows the polarization measured for a very thin Co film backed by
a normal metal film to provide lateral conductivity [7]. Half the bulk
value of P is reached for an average thickness of Co equal to about two
atomic layers. Since the method can detect changes in P of 0.1% with the
present technique and this could easily be improved, we could see a very
small fraction of a monolayer of a ferromagnetic film.

The explanation of the ferromagnetic results is still in question but a
consistent picture seems to be developing. The basic idea is that it is the
almost free electrons that contribute to the tunnel currents and the large

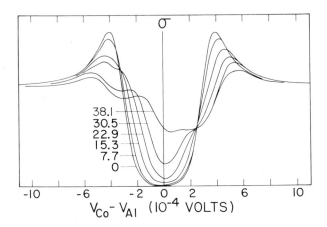

Fig.6 Tunneling conductance as a function of applied voltage for an
Al/Al$_2$O$_3$/Co tunnel junction in various parallel magnetic fields

234

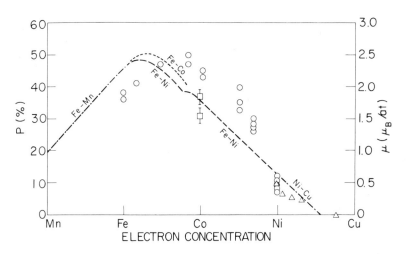

Fig.7 The electron spin polarization as determined by electron tunneling into various ferromagnetic elements and alloys as a function of electron concentration is given by the points. The broken lines show the saturation magnetic moment (right hand scale) according to the Slater-Pauling curve

density of localized d states at the Fermi energy in the transition metal ferromagnets contribute little. Theoretical contributions by HERTZ and AOI [9] and by CHAZALVIEL and YAFET [10] have emphasized this point and also the importance of s-d hybridization. In the limit of almost free electrons these analyses reduce to the simple assumption of STEARNS [11] that the tunneling currents are proportional to the spin density of states of the almost free electrons at E_F. Such a model gives the correct sign of P, the

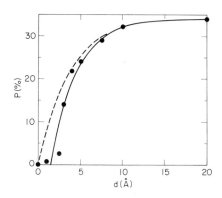

Fig.8 Electron spin polarization of tunneling currents from ultra-thin Co films showing that one half of the thick film value is reached by an average film thickness of about 2 atomic layers

approximate magnitude, and explains the proportionality to the magnetic moment.

Recently we have measured the polarization of the tunnel currents from the rare earth ferromagnets. In some ways the ferromagnetism in these elements is simpler than in the transition metals. The 4f electrons which contribute most of the magnetic moment are completely localized and definitely do not tunnel. The ferromagnetism coupling between the ion cores is by way of the conduction electrons, which thereby become positively polarized and contribute a little to the total magnetic moment. In this case we know that the average magnetic moment of the conduction electrons is positive and that the tunneling currents are also positively polarized. This reinforces the interpretation of the transition metal ferromagnets. For instance for Gd the total moment is +7.55 Bohr magnetons of which +0.55 is attributed to 3 conduction electrons. The measured polarization $P \approx + 15\%$.

I will just mention very briefly how this technique might be applied to inelastic tunneling. Magnons have already been seen [12] in the low energy region and could be studied by this method. The matter of tunneling anomalies has been covered elsewhere, but the use of spin polarization might contribute to the understanding. One variation of the present technique of studying vibrational spectra would be to measure the spectra with tunnel currents with oriented spins. An inelastic collision of the spin polarized electron with an ion with an oriented spin which is part of a molecule will result in exciting the vibrational modes of the molecule. The strength of absorption will presumably depend on the relative orientation of the spins. Another possible application would be in the study of electronic states such as those being studied by LEGER et al. [13].

I will end by pointing out some of the strengths and drawbacks of the sources of spin-polarized electrons. Using a superconductor gives very high resolution. Unfortunately for most vibrational spectra the two spin-polarized electron currents can only be separated in energy about equal to the superconducting energy gap which for the present Al films is about 1 meV. It is possible that this might be increased to about 5 meV separation for high temperature superconductors. However, unless the vibrational level being investigated is narrower than the splitting of the sources, the spectrum will not be noticeably different from an unpolarized source as far as line position is concerned. The experiments to date have been done at liquid helium three temperature (0.4K) which is not very convenient. It is possible that other superconductors with a high T_c can replace Al as a spin-polarized source and allow the measurements to be made at liquid helium 4 temperatures. One possibility is Be if the phase whose transition temperature is 8 K can be stabilized [14]. It could be used at 1.5 K or even 4.2 K, would allow a high magnetic field (150 kG) and therefore more spin splitting and it would have a much lower spin-orbit scattering than Al because of its low atomic number, which would simplify the analysis. Possibly some type II superconductors would also be useful, although the spin-orbit scattering will be worse than in Al. It should be pointed out that in working with superconducting films the alignment of the plane of the film with the magnetic field to within about 0.1 degree is very important.

Another possible method is to use a ferromagnet as a source of spin polarized electrons. Fe-Ni alloys can have a 50% polarization and field emission experiments have shown that with Eu S_2 90% polarization can be attained. Such experiments have not been done in tunneling, but they may be feasible. The resolution attainable will be determined by the operating temperature which no longer needs to be very low. In such a case the magnetic field needed would be only that required to align the domains of the ferromagnet and the spins could be oriented in any direction. A junction could be made of Al-Al$_2$O$_3$- organic material -Fe except that the blanketing effect of Fe on the organic material would be much worse than Pb. However, it might be possible to put on a very small amount of Fe and then a thicker lead blanket. None of these things have been tried but some of the techniques are workable if one really has a problem to be solved with spin dependent inelastic tunneling.

References

1 R. Meservey, P.M. Tedrow: J. Appl. Phys. 42, 51 (1971)
2 R. Meservey, P.M. Tedrow, P. Fulde: Phys. Rev. Lett. 25, 1270 (1970)
3 H. Engler, P. Fulde: Z. Physik 247, 1 (1971)
4 For a review of much of the work on spin paramagnetism in superconductors see P. Fulde: Advances in Physics 22, 667 (1973)
5 R.C. Bruno, B.B. Schwartz: Phys. Rev. B8, 3161 (1973). See also R. Meservey, P.M. Tedrow, R.C. Bruno: Phys. Rev. B11, 4224 (1975)
6 P.M. Tedrow, R. Meservey: Phys. Rev. B7, 318 (1973)
7 R. Meservey, D. Paraskevopoulos, P.M. Tedrow: Phys. Rev. Lett. 37, 858 (1976)
8 P.M. Tedrow, R. Meservey: Solid State Comm. 16, 71 (1975)
9 J.N. Chazalviel, Y. Yafet: Phys. Rev. B 15, 1062 (1977)
10 J.A. Hertz, K. Aoi: Phys. Rev. B 8, 3252 (1973)
11 M.B. Stearns: J. Mag. and Mag. Mat. 5, 167 (1977)
12 D.C. Tsui, R.E. Dietz, L.R. Walker: Phys. Rev. Lett. 27, 1729 (1971)
13 A. Léger, J. Klein, B. Balin, D. Defourneau: Solid State Comm. 11, 1331 (1972)
14 P.M. Tedrow, R. Meservey: Phys. Lett. 58 A, 237 (1976)
15 N. Müller, W. Eckstein, W. Heiland: Phys. Rev. Lett. 29, 1651 (1972)

Semiconductor Tunneling, Bound Levels and Band Structure

D.C. Tsui

Bell Telephone Laboratories
Murray Hill, NJ 07974, USA

ABSTRACT

Bound levels due to quantum size effects in solids exist in a variety of
physical systems. These include space-charge layers at semiconductor surfaces, thin films, and thin films bounded by solid-solid interfaces such as
the normal metal-superconductor interface, the semiconductor-semiconductor
interface, or the semiconductor-insulator interface. The discovery of such
bound levels and the determination of their splitting were in most cases
accomplished by utilizing the electron tunneling technique. This talk gives
a survey of the experimental studies of bound levels and the related band
structure effects by electron tunneling.

Organization

Conference on Inelastic Electron Tunneling Spectroscopy and Symposium on Electron Tunneling.

<u>Organizing and Program Committee</u>

Chairman: T. Wolfram, Dept. of Physics, University of Missouri-Columbia
Members: R.V. Coleman, Dept. of Physics, University of Virginia
          P.K. Hansma, Dept. of Physics, University of California, Santa
              Barbara
          R.C. Jaklevic, Research Staff, Ford Motor Company
          J. Lambe, Research Staff, Ford Motor Company
          J. Rowell, Bell Laboratories
          D.J. Scalapino, Dept. of Physics, University of California,
              Santa Barbara
          W.H. Weinberg, Dept. of Chemical Engineering, California Institute
              of Technology
          A. Yelon, Dept. génie Physique, Ecole Polytechnique, Montréal

<u>Logistics and Arrangements</u>

          University of Missouri-Columbia, Extension Division
          Mrs. Beverly Huckaba
          Mrs. Bonnie Beckett

Conference on Inelastic Electron Tunneling Spectroscopy

PROGRAM

MAY 25, 1977

WELCOME: Dr. James C. Olson, President
          University of Missouri

OPENING REMARKS: T. Wolfram, University of Missouri
                  Department of Physics

SESSION CA: REVIEW OF INELASTIC ELECTRON TUNNELING
             W.J. Tomasch, Chairman
             Department of Physics, University of Notre Dame

CA1: Inelastic Electron Tunneling Spectroscopy - History and Future
      R.C. Jaklevic, Research Staff, Ford Motor Company

CA2: Survey of Applications of Tunneling Spectroscopy
      P.K. Hansma, Department of Physics, University of California, Santa
      Barbara

CA3: Theoretical Interpretation of IETS Data
      J. Kirtley, Department of Physics and Laboratory of Matter, University
      of Pennsylvania

SESSION CB: APPLICATIONS OF INELASTIC ELECTRON TUNNELING
             J.G. Adler, Chairman
             University of Alberta

CB1: Application of IETS to Surface Chemistry
      W.H. Weinberg, Department of Chemical Engineering, California Institute
      of Technology

CB2: Application of IETS to the Study of Biological Materials
      R.V. Coleman, Department of Physics, University of Virginia

CB3: Application of IETS to Trace Substance Detection
      A. Yelon, Department of génie Physique, Ecole Polytechnique, Montréal

CD3: Relation of IETS to Other Surface Studies
      Opening comments by W. Plummer, Department of Physics, University of
      Pennsylvania

CD4: Effects of Cooperative Behavior on Molecular Vibrational IETS Peak
      Intensities
      S.L. Cunningham, California Institute of Technology

CD5: Phonon Damping of Virtual Levels in Thick Superconducting Film
W.J. Tomasch, Department of Physics, University of Notre Dame

CD6: The Golden Rule Formalism in Tunneling - Can it be Justified?
T.E. Feuchtwang, Osmond Lab, Pennsylvania State University

MAY 26, 1977

SESSION CE: MOLECULAR ADSORPTION ON NON-METALLIC SURFACES
H. Jarrett, Chairman
E. I. DuPont Company

CE1: Photoemission Studies of Molecular Adsorption on Oxide Surfaces
V.E. Henrich, Lincoln Laboratory, Massachusetts Institute of Technology

CE2: Calculation of the Orientational Dependence of IETS Intensities for Ethylene
J. Rath, Department of Physics, University of Missouri-Columbia

CE3: Structure and Dynamics of Butane Adsorbed on Graphite by Inelastic Neutron Scattering
H. Taub, Department of Physics, University of Missouri-Columbia

CB4: Application of IETS to the Study of Adhesion
H. White, Department of Physics, University of Missouri-Columbia

MAY 26, 1977

SESSION CC: THEORY: INELASTIC ELECTRON TUNNELING-SCATTERING AT SURFACES
D.J. Scalapino, Chairman
University of California - Santa Barbara

INTRODUCTORY REMARKS: D.J. Scalapino

CC1: Calculations of Inelastic Tunneling Cross Sections Using Self-Consistent Multiple Scattering Techniques
J.R. Schrieffer, Department of Physics, University of Pennsylvania

CC2: Interaction of Low Energy Electron Beams with Surface Vibrations
D.L. Mills, Department of Physics, University of California - Irvine

CC3: Theory of Surface Plasmon Excitations by Electron Tunneling
L.C. Davis, Research Staff, Ford Motor Company

SESSION CD: PANEL DISCUSSIONS AND SELECTED PAPERS

CD1: Technology of IETS
Opening comments by J.G. Adler
Department of Physics, University of Alberta

CD2: Problems in the Biological Sciences
Opening comments by L. Sherman
Division of Biological Sciences, University of Missouri-Columbia

Symposium on Electron Tunneling

Interactions on Metal Surfaces

Editor: R. Gomer

1975. 112 figures. XI, 310 pages
(Topics in Applied Physics, Volume 4)
ISBN 3-540-07094-X

Contents:
J.R. Smith: Theory of Electronic Properties of Surfaces
S.K. Lyo, R. Gomer: Theory of Chemisorption
L.D. Schmidt: Chemisorption: Aspects of the Experimental Situation
D. Menzel: Desorption Phenomena
E.W. Plummer: Photoemission and Field Emission Spectroscopy
E. Bauer: Low Energy Electron Diffraction (LEED) and Auger Methods
M. Boudart: Concepts in Heterogeneous Catalysis

"All in all this volume is a timely, well written and professional review of surface physics. Clearly there are many areas of surface science which are not discussed, however, the editor has wisely focused on those areas of high current interest."

J.R. Schrieffer in: Applied Physics

"...The book gives a clear view of recent progress in the selected branches of surface physics. The different articles are comprehensive and clearly written. The book is valuable for scientists working in this field as well as for those wanting to be informed about the current state of our understanding of surface phenomena."

F. Baumann in: Journal of Nuclear Materials

Electron Spectroscopy for Surface Analysis

Editor: H. Ibach

1977. 123 figures, 5 tables. XI, 255 pages
(Topics in Current Physics, Volume 4)
ISBN 3-540-08078-3

Contents:
H. Ibach: Introduction
D. Roy, J.D. Carette: Design of Electron Spectrometers for Surface Analysis
J. Kirschner: Electron-Excited Core Level Spectroscopies
M. Henzler: Electron Diffraction and Surface Defect Structure
B. Feuerbacher, B. Fitton: Photoemission Spectroscopy
H. Froitzheim: Electron Energy Loss Spectroscopy

The great progress made in the modern understanding of surfaces and interfaces is chiefly due to the technique of electron spectroscopy. Consequently, electron spectroscopies are no longer instruments of solely fundamental research, but can be found today even in industrial laboratories. The more practically oriented scientists and engineers may be bewildered by the number of different electron spectroscopies that have become known under their various acronyms. This book specifically addresses the question of which information is best obtained by a certain type of spectroscopy and which combination of spectroscopies might be most suitable for a certain application. Theoretical considerations are kept brief for the benefit of a more extended presentation of experimental aspects and examples.

Springer-Verlag
Berlin Heidelberg New York

Applied Physics

A monthly journal

Board of Editors	**S. Amelinckx,** Mol. · **V. P. Chebotayev,** Novosibirsk
	R. Gomer, Chicago, Ill. · **H. Ibach,** Jülich
	V. S. Letokhov, Moskau · **H. K. V. Lotsch,** Heidelberg
	H. J. Queisser, Stuttgart · **F. P. Schäfer,** Göttingen
	A. Seeger, Stuttgart · **K. Shimoda,** Tokyo
	T. Tamir, Brooklyn, N.Y. · **W. T. Welford,** London
	H. P. J. Wijn, Eindhoven

Coverage

application-oriented experimental and theoretical physics:

Solid-State Physics	*Quantum Electronics*
Surface Physics	*Laser Spectroscopy*
Chemisorption	*Photophysical Chemistry*
Microwave Acoustics	*Optical Physics*
Electrophysics	*Integrated Optics*

Special Features

rapid publication (3–4 months)
no page charge for **concise** reports
prepublication of titles and abstracts
microfiche edition available as well

Languages

Mostly English

Articles

original reports, and short communications
review and/or tutorial papers

Manuscripts

to Springer-Verlag (Attn. H. Lotsch), P.O. Box 105 280
D-69 Heidelberg 1, F.R. Germany

Place North-American orders with:
Springer-Verlag New York Inc., 175 Fifth Avenue, New York. N.Y. 10010, USA

Springer-Verlag
Berlin Heidelberg New York